Protecting human rights in a new South Africa

Albie Sachs

CONTEMPORARY SOUTH AFRICAN DEBATES

Oxford University Press
Cape Town

Oxford University Press
Walton Street, Oxford OX2 6DP, United Kingdom

Oxford New York Toronto
Delhi Bombay Calcutta Madras Karachi
Petaling Jaya Singapore Hong Kong Tokyo
Nairobi Dar es Salaam Cape Town
Melbourne Auckland

and associated companies in
Berlin Ibadan

ISBN 0 19 570 609 9

© Albie Sachs 1990

Oxford is a trademark of Oxford University Press

All rights reserved. No part of this publication may be reproduced, stored in a retrieval system, or transmitted in any form or by any means, electronic, mechanical, photocopying, recording or otherwise, without the prior permission of the copyright owner.

Published by Oxford University Press Southern Africa
Harrington House, Barrack Street, Cape Town, 8001, South Africa

DTP conversion by CAPS of Cape Town in 10 on 12 pt Garamond
Printed and bound by Clyson Press, Maitland, Cape

Contents

About the author   *iv*
Acknowledgements   *vi*
Preface   *vii*

1   Towards a Bill of Rights for a democratic South Africa   1
2   Evolving a Bill of Rights culture   32
3   To believe or not to believe   43
4   Free speech — limited or qualified   50
5   Judges and gender: The constitutional rights of women in a post-apartheid South Africa   53
6   The constitutional position of the family in a democratic South Africa   64
7   Towards a charter of children's rights   79
8   The future of South African law   90
9   Rights to the land   104
10   Conservation and third generation rights: The right to beauty   139
11   The constitutional position of whites in a democratic South Africa   149
12   Preparing ourselves for freedom: Culture and the ANC Constitutional Guidelines   175
13   The last word — freedom   184

Appendix 1: Two underlying questions   193
Appendix 2: The ANC's Constitutional Guidelines for a democratic South Africa   197

About the author

Albie Sachs was born in Johannesburg in 1935 and schooled at the South African College High School (SACS) in Cape Town. He obtained a BA LL B from the University of Cape Town and then practised as an advocate in the Supreme Court of South Africa, Cape Town.

After twice being detained by the Security Police, Albie Sachs went to England on an exit permit. There he took a Ph.D. at Sussex University. He later became senior lecturer in law at the University of Southampton, and received an Honourary LL D from this university.

From 1977 until 1983 he was Professor of Law at Eduardo Mondlane University in Mozambique and was Director of Research in the Mozambican Ministry of Justice. In 1988 he survived a car bomb attempt on his life.

Since then he has taught at Columbia University in New York, and is at present Director of the South African Constitution Studies Centre, Institute for Commonwealth Studies at the University of London. He is also presently attached to the University of the Western Cape and the University of Cape Town.

Albie Sachs is a member of the African National Congress' Constitutional Committee, but this book has been written in his personal capacity.

Albie Sachs
© Rashid Lombard, 1990

Acknowledgements

The Publishers would like to thank the Institute for a Democratic Alternative for South Africa (IDASA) for assisting with the production costs.

Preface

If a constitution is the autobiography of a nation, then we are the privileged generation that will do the writing. It is something that involves us all. Our country, our future, our rights are at stake.

No one gives us rights. We win them in struggle. They exist in our hearts before they exist on paper. Yet intellectual struggle is one of the most important areas of the battle for rights. It is through concepts that we link our dreams to the acts of daily life.

We are not used to the idea of rights, certainly not of constitutional rights. Our debates are about power rather than rights. We speak about human rights only in terms of how they are violated and not in terms of how they can affirm and legitimize a new society. If we can agree upon the basic rights, freedoms, and relationships we want in a new South Africa, then the question of formulating precise governmental structures and electoral procedures will not be difficult.

Without a clear and vigorous concept of rights, non-racial democracy is like a fountain without water, beautiful but stony. We must give texture and flow to non-racial democracy. Much suffering and pain have gone into its achievement. It is the basis for unifying the nation, and the context for the expression of our political rights. Yet it does not in itself solve the question of reconciling equality with cultural diversity. It does not tell us how to harmonize rights of individual liberty with rights of social progress. It does not answer the question of what principles should govern the sharing of the land. It does not address the issue of gender rights, of workers' rights, of children's rights, of whether there should be limits to freedom of speech in order to avoid racial explosions, of how simultaneously to guarantee the rights of the people as a whole while allaying the fears of those who regard themselves as a minority.

What follows is an attempt to apply the logic of a human rights approach to the building of non-racial democracy in South Africa. Some of the formulations may be of my invention, but the ideas have all emerged from struggle and it is my hope that through these pages they will return to their source.

Basic to the approach I have developed is the idea that the general population should be involved as far as possible in formulating the rights they wish to enjoy. Ideally this should be done before a new constitution is adopted. The need to put democratic political institutions in place is so strong, however, that the rights debate and the evolution of specific charters of rights is likely to continue well into the post-apartheid era.

Preface

Readers will have no difficulty in discerning the influence of decades of involvement with the ANC, starting with participation in the Defiance of Unjust Laws Campaign in the early 1950s and culminating in membership today of the ANC's Constitutional Committee. Those who know about these things will also immediately understand that what follows is not an official presentation of ANC views, nor even an unofficial one, but a small personal contribution to the great national debate which the ANC has done so much to encourage.

I would like to dedicate this book to Oliver Tambo. One day the story will be told of the contribution he has made and is still making to the creation of a new South Africa and the influence he has had on all of us. If ever there was a democrat, a patriot and a lover of freedom, it is he.

Albie Sachs
London and Cape Town
August 1990

1 Towards a Bill of Rights for a democratic South Africa

All revolutions are impossible until they happen; then they become inevitable. South Africa has for long been trembling between the impossible and the inevitable, and it is in this singularly unstable situation that the question of human rights and the basics of government in post-apartheid society demands attention.

No longer is it necessary to spend much time analysing schemes to modernize, reform, liberalize, privatize, or even democratize apartheid. Like slavery and colonialism, apartheid is regarded as irremediably bad. There cannot be good apartheid, or degrees of acceptable apartheid. The only questions are how to end the system as rapidly as possible and how to ensure that the new society which replaces it lives up to the ideals of the South African people and the world community. More specifically, at the constitutional level, the issue is no longer whether to have democracy and equal rights, but how fully to achieve these principles and how to ensure that within the overall democratic scheme, the cultural diversity of the country is accommodated and the individual rights of citizens respected.

Five constitutional schemes

Proposals for new constitutional dispensations have been made in the past decade on almost a monthly basis. South African air has been thick with a specially invented or adapted vocabulary: confederation, consociation, tricameral, three-tier, group rights, own affairs. Behind the multiplicity of reports and proposals, however, it has been possible to discern a number of major positions. For the sake of convenience, and bearing in mind that the categories shade into each other, five basic constitutional schemes may be distinguished. They can be summarized as follows:

☐ open apartheid;
☐ reformed apartheid;
☐ multiracial apartheid;
☐ hidden or democratic apartheid; and
☐ anti-apartheid (non-racial democracy).

The terminology is, of course, not that of the authors of the proposals since most of them insist that their schemes would end rather than perpetuate apartheid. But what the first four proposals have in common is that they have all been based on a desire to preserve a constitutionally privileged position for the white minority. Furthermore, all four proposals have made the distribution of power and wealth depend on the criterion of race, the first three directly, the fourth indirectly.

Open apartheid

The basic constitutional tenets of open apartheid are well known. They presuppose separate sovereignties for whites and blacks with no constitutional mixing at any level. In the original formulation, whites were to have exclusive control over so-called white South Africa, that is, eighty-seven per cent of the surface area of the country, including all the developed zones. Blacks were to become independent in their so-called tribal homelands. Even blacks living in the so-called white areas were to exercise their rights through the bantustans to which they were attached by descent and language. Ethnicity was given a territorial base and was made the exclusive constitutional principle. Relations between black and white became relations of international law, not of constitutional law.

Today the supporters of open apartheid are just a tiny bit more modest. They demand a separate white or Afrikaner state, but cannot agree on where it should be.

Reformed apartheid

Reformed apartheid sought to make race the dominant but not the exclusive principle of the constitution. It based political rights on race but recognized that some sort of political interrelationship involving all ethnic groups was necessary. The term most frequently used was confederation. Essentially it presupposed links between the white-dominated central Parliament and the bantustans. To complete the picture, South Africans of mixed or Indian descent (almost completely ignored in the open apartheid scheme) were to be junior partners in the tricameral Parliament, and so-called urban blacks were to have a series of councils, starting at the community level and moving upwards, to represent their interests. Apartheid would remain intact in that all organs of legislative power would continue to be established on separate ethnic bases, each one having exclusive control over what was defined as 'own affairs'. The element of reform was to be contained in a provision that 'common affairs' would be dealt with at a high level on the basis of meetings between representatives of the different groups in some form of confederal council. Since everybody would have the vote at some level or other, it would be claimed that the principle of universal suffrage was being recognized. At the same time, overtly discriminatory laws would be gradually reduced.

A fundamental feature of this scheme was that through dividing the black population, through regulating numbers at crucial levels, through the definition of own affairs and common or general affairs, through the control of funds, and control of the state apparatus including the army and police force, the white minority and specifically the National Party would maintain control over the country. This would be a form of limited power-sharing under the clear hegemony of the leading party in the white chamber of the tricameral Parliament.

It was the manifest failure of this scheme that produced the crisis leading to recognition of the central role of the ANC and of the need for serious negotiations.

Multiracial apartheid

In essence multiracial apartheid, which is still on the agenda but for which support is fading, is based on the politics of inter-ethnic alliance rather than inter-ethnic consultation. The bantustans or tribal homelands retain some importance but are not projected as the sole structures through which Africans can exercise political rights. Rather, they are gradually integrated as component parts of regional political structures, retaining some autonomy, but sharing certain powers on a regional basis with the people living in the so-called multiracial areas.

The foundation of this approach lies in the report of a Commission Chief Gatsha Buthelezi set up some years back to inquire into the future of the province of Natal. More recently, the report evolved into the so-called KwaZulu-Natal Indaba proposals. The region is projected as the embryo unit of a future federal state. Regions may have differences in their political structures and advance at different rates. The federal government is gradually structured on the basis of interaction between the leaders of the regions. The way is left open for a black Head of State, who by virtue of his/her own election to office will declare that apartheid is dead and buried. What are referred to as legitimate white fears are constitutionally catered for by means of a combination of territorial divisions, own affairs concepts, racial quotas in government, group vetoes for minorities, privatized racism through voluntary association, and entrenched group and individual rights. Behind all these devices are two fundamental principles: there shall be no majority rule and there shall be no rapid moves to end the inequalities produced by apartheid.

Recently, Pretoria has begun more and more to adopt these positions, though now it uses the language of minority rights rather than group rights.

Hidden or democratic apartheid

Hidden or democratic apartheid starts off on the democratic assumption, but is reluctant to accept universal suffrage in a unitary state. It accepts the hypothesis that the African National Congress (ANC) would probably be the ruling party in the new society (ours being the only revolution to be accompanied by opinion polls). Where the apartheid aspect would live on buried in the heart of the new democratic constitution would be in entrenched clauses which would be insisted upon as the condition for the acceptance of the principle of one person, one vote. Such clauses would impose a double brake on the dismantling of apartheid; they would gravely restrict the competence of Parliament, especially in economic matters, and they would institutionalize conservative and white-dominated machinery to guarantee that such competence is narrowly interpreted. Thus, they would abolish *de jure* apartheid but constitutionally freeze the existing *de facto* apartheid whereby eighty-seven per cent of the land and ninety-five per cent of the country's productive capacity belong to whites.

Non-racial democracy

Non-racial democracy presupposes a united South Africa governed by the principles of universal suffrage, majority rule, and equal individual rights. The Freedom Charter, adopted by the Congress of the People in 1955, sets out a clear programme born out of South African reality which for many of us would serve as the fundamental document around which a new constitution could be developed. Others might arrive at similar positions by different routes. But within the basic framework of the rights and freedoms set out in the Freedom Charter, and with a view to making its principles the property of all the South African people, there would be many issues which could be discussed: the internal structure of the government, whether to have a Presidential or Prime Ministerial form of leadership, what the official languages should be, the role of an upper house, the electoral system, the territorial divisions of the country, and where the capital should be situated. Perhaps more important, negotiations could play a key role in providing for the orderly transfer of power from a racial minority to all the people. Once the principle is accepted that apartheid is to be completely dismantled, and once it is agreed that the only effective and lasting way to dismantle it is to establish a non-racial, non-sexist, democratic society in a united country, the issue of how to proceed most rapidly to the materialization of this solution comes to be placed squarely on the agenda.

The ANC has in recent years opted for a Bill of Rights enforceable through the courts and has accepted that there will be a mixed economy in which the state will play an important role. On the other hand, supporters of the 'hidden/democratic apartheid' option, mentioned above, are moving closer to the idea that there has to be some economic redistribution. Thus the gap between these two options is narrowing. Former supporters of the multiracial apartheid option are also beginning to accept that non-racial democracy offers a far more secure future for whites, as for all South Africans, than would any attempt to entrench group rights. The possibilities of a democratically based consensus are far stronger than a few years ago; and as the revolution becomes increasingly 'more inevitable' and increasingly 'less impossible', so do the chances of a peaceful constitutional resolution improve.

A Bill of Rights for a post-apartheid South Africa: some misconceptions

Two widely held but opposing views on a Bill of Rights argue in summary that either

- a Bill of Rights is necessary because if you grant the legitimate rights of the black majority you must also give reasonable protection to the rights of the white minority, or

- [ ] a Bill of Rights is a reactionary device designed to preserve the interests of whites and to prevent any effective redistribution of wealth and power in South Africa.

The most curious feature about the demand for a Bill of Rights in South Africa is that initially it came not from the ranks of the oppressed but from a certain stratum in the ranks of the oppressors. This had the effect of turning the debate on a Bill of Rights inside out. Instead of being welcomed by the mass of the population as an instrument of liberation, it was viewed by the majority with almost total suspicion as a brake on advance. Indeed, South Africa must be the only country in the world in which sections of the oppressed actually constituted an anti-Bill of Rights Committee.

At first sight, nothing would appear simpler than to adopt a Bill of Rights based upon a universally accepted document such as the United Nations Universal Declaration of Human Rights. The fact is that the apartheid divide lies as heavily on the Bill of Rights debate as it does on any other important topic in South Africa. Disagreement relates not only to the specific clauses to be included or excluded, but to the whole thrust of a possible Bill, to the manner in which it should be created, and to the means whereby it should be enforced.

Suspicions about a Bill of Rights

It is a sad tribute to the way the law has impinged on the life of the majority of South Africans that a Bill of Rights has been seen essentially as a means of using juridical techniques to restrict rather than enlarge the area of human freedom. Suspicion has been founded on a variety of interconnected factors:

- [ ] The push for a Bill of Rights came not from the heart of the freedom struggle, but from people on the fringes, many of whom have criticized apartheid, but few of whom have been actively involved in the struggle against it.

- [ ] The objective of the Bill of Rights was seen as being primarily to protect the existing and unjustly acquired rights of the racist minority rather than to advance the legitimate claims of the oppressed majority.

- [ ] The attack on majoritarianism, which underlay many arguments in favour of a Bill of Rights, was manifestly racist, since South Africa has been governed without a Bill of Rights and in accordance with the principles of majority rule (for the minority!) since the Union of South Africa was created in 1910. It is only now that the majority promises to be black, that constitutional doubts and the need for checks and balances suddenly become allegedly self-evident.

☐ The key role given to what are called experienced lawyers in controlling the implementation of the proposed Bill of Rights would have meant inevitably an interpretation in favour of the existing and unjustly acquired rights and against any meaningful re-allocation of rights.

☐ While protection of the individual was accepted as necessary, the failure of the proposed Bill of Rights to address the question of grossly disadvantaged communities rendered it largely irrelevant to the human rights needs of the country.

Such suspicions might have seemed shockingly unjust to proponents of a Bill of Rights, many of whom had both a genuine hatred of apartheid and a deeply sincere desire to see as rapid and peaceful a transformation of the country as possible. Yet the proponents of a Bill of Rights have rushed ahead with their drafts without paying due attention to questions to which their lawyerly background should have made them more sensitive.

Before drafting a Bill of Rights for a post-apartheid South Africa, it is necessary to ask certain preliminary questions, the answers to which will decisively affect the final result. More specifically, it is necessary to ask simply:

What Bill of Rights?
By whom and for whom?
How?

Misconceptions about the content of a Bill of Rights — the question of the three generations of rights

Most proponents of a Bill of Rights for South Africa operate within a thematically limited and historically out of date perspective. Very few get beyond what has been called the first generation of rights, namely, civil and political rights and rights of due process, as were declared during the great anti-feudal and anti-colonial revolutions of the eighteenth century. The second generation of rights, namely those of a social, economic, and cultural nature enunciated in the United Nations Charter of Human Rights of the 1960s, get scant mention. The third generation of rights, the rights to development, peace, social identity, and a clean environment, which have been clearly identified as human and people's rights only in the past decade, get virtually no attention at all. At a time when every possible intellectual input is needed, it is perverse indeed to restrict the scope of the debate to first generation rights only, just as it would be grossly anachronistic to start post-apartheid South Africa with a Bill of Rights document as archaic (even if not as vicious) as the system it is designed to eliminate.

The great majority of South Africans have in reality never enjoyed either first, second, or third generation rights. Their franchise rights have been restricted or non-existent, so the achievement of first generation rights is fundamental to the establishment of democracy and the overcoming of national oppression. But for the vote to have meaning, for the Rule of Law to have content, the vote must be the instrument for the achievement of second and third generation rights. It would be a sad victory if the people had the right every five or so years to emerge from their forced-removal hovels and second-rate Group Area homesteads to go to the polls, only thereafter to return to their inferior houses, inferior education, and inferior jobs. Indeed it would be a strange panoply of rights that not only ignored but excluded the right to peace and development, the right to enjoy the beauty of and benefit from the natural resources of the country, and above all, the right to be a people, to be South African in the full sense of the word, to constitute a nation, to overcome the divisions and inferiorization imposed by racism, tribalism, and regionalism, and to participate as a people in the life of the community of nations.

There are some persons who would wish to restrict the extension of rights to the first generation only, granting formal political power but depriving it of practical content; the people can have the vote, but not homes and jobs. There are others who would see the extension of second generation socio-economic rights as an alternative to first generation political rights; the people can have homes and jobs, but not the vote. Very few look at the third generation at all, the rights so important to a people denied peace, security, dignity, and an identity for centuries.

The fundamental constitutional problem, however, is not to set one generation of rights against another, but to harmonize all three. The possessors of the rights are the same physical human beings, namely, the citizens of a democratic South Africa. They do not exercise one set of rights in the morning, another in the afternoon, and a third at night. The web of rights is unbroken in fabric, simultaneous in operation, and all-extensive in character. In the world at large, the generations of rights, or rather, of rights formulations, succeeded each other at different times, but their sphere and object was essentially the same and their line of development has been continuous. It would be absurd for us in South Africa to have to recapitulate and live through each stage separately before advancing to the next. We do not need to re-invent each formulation. Rather, we draw from the intellectual creation of the world community to find formulae and solutions to our own problems. Thus, when the majority in South Africa look to the complete elimination of apartheid in all its shapes and forms, what they are longing for is the progressive, rapid, and simultaneous achievement of all the rights as formulated in all three generations. The people of South Africa want to be free, to live decent lives, to be a community with their own personality and culture, and to live in peace and with dignity with each other and the world, no more, no less.

In a phrase, they wish to exercise simultaneously what Kader Asmal, member of the ANC Constitutional Committee, has felicitously called blue rights, red rights, and green rights. Each has its own sphere, its own modalities of enforcement; each has a fundamental and irreducible character, but all need to be taken together in framing a new constitution.

A Bill of Rights: by whom and for whom?

A look at the historical contexts in which other Bills of Rights have been adopted shows the back-to-front nature of the proposed Bill of Rights for South Africa. The Magna Carta, the charter of rights adopted in the seventeenth century, the United States Bill of Rights, and the French Declaration of the Rights of Man were all formulated and adopted by the former victims of arbitrariness and oppression as a means of controlling or excluding the power of the former oppressors and guaranteeing the aggrieved classes freedom from future revivals of arbitrary behaviour. It was not Hitler or his former supporters who drafted the United Nations Declaration of Human Rights or the subsequent charter, but rather the survivors of the Hitlerite pillage and massacre, supported by a shocked world.

If we take a close look at the great prototype Bill of Rights, namely that contained in the early amendments to the United States Constitution, we see that it was adopted not by the ousted colonial authorities but by the victorious freedom-fighters. We observe too that the objective of the amendments was not to preserve the rights of the defeated loyalists, but rather to root out once and for all the kinds of oppressive behaviour indulged in by supporters of the Crown. Thus, each of the amendments was designed to deal with a specific form of denial of rights: no freedom of speech of assembly, the imposition of an official church, the use of torture and cruel punishments, the forcing of confessions, and so on. The Bill was not an abstractly conceived set of rights designed by lawyers in terms of general, pre-conceived notions, but a concrete set of responses to specifically felt forms of domination. The former colonized people, victims of despotism, anxious to guarantee that there be no revival of the suffering to which they had been subjected, and to consolidate their new-found freedom, remembered exactly where the shoe had pinched, and designed their Bill of Rights to respond accordingly.

Applied to South Africa, this would mean essentially that the Bill of Rights would be adopted after freedom had been won, and as a means of ensuring that oppression was not restored in old or new forms. The Bill of Rights would confront and outlaw all the specific forms of oppression associated with apartheid: the whole system of racial domination, the pass laws, legally enforced removals, the Group Areas legislation, and the violence of the troops in the townships and of the security police in the cells. Since the equivalent of independence in South African conditions is the restoration of the land, of dignity, and rights within the existing boundaries of the country, a Bill of Rights would have to address itself directly to the question of equal access to resources. In other words, a genuine document in the classic Bill of Rights tradition would have as its principal objective the total elimination of apartheid and the guaranteeing of rights to those presently oppressed. In attending to these issues it would speak out in general language guaranteeing rights to the whole population.

In the proposals being made we find almost exactly the opposite expressed. The principle objective is precisely to give guarantees to the present oppressors, to protect them against the re-vindications of the oppressed; to do so in advance of and as a bulwark against rather than as a prescription in favour of change. Such a Bill of Rights would be deprived of its true function. Instead of being an instrument designed to protect the future rights of the whole population, it becomes a means of defending the present privileges of the minority, surpassing the legitimate bounds of legal irony by making a Bill of Rights the document that perpetuates injustice. It is only necessary to refer to a concrete example to see the significance of this issue.

If one looks at the question of the land, one sees the contradiction immediately. In the past three decades, more than three million South Africans have been forcibly removed from their homes and farms, simply because they were black. Apartheid law then conferred legal title on owners whose main merit was that of having a white skin. Whom would the proposed Bill of Rights protect: the victims of this unjust conduct, which has been condemned as a crime against humanity by all humankind, or the beneficiaries? Although oppression and poverty are not necessarily completely synonymous, they do tend to go hand in hand. Where would the people, condemned as squatters after having living on land for generations, their homes bulldozed into the ground, get the finance to compensate the new owners who have legal 'titles', when the only collateral the uprooted would have has no known market value, namely, centuries of suffering and dispossession?

Looking at the surface area of South Africa as a whole, one finds that at present the dominant minority of less than twenty per cent of the population has reserved to them by law nearly ninety per cent of the land. It would be a strange Bill of Rights that said in effect that the remaining eighty per cent of the population had to forego their right to own and farm land because to exercise such a right would be to violate the acquired apartheid rights of the twenty per cent. Looked at from the perspective of human rights, who has the greater claim to land — the original owners and workers of the land, expelled by guns, torches, and bulldozers from the soil, turned into migrant workers, perpetually on the move with no plot they can call their own; or the present owners, frequently absentee, whose rights are based on titles conferred in terms of the so-called Native Lands Act and the Group Areas Act?

This is not to say that there are no white farmers with a deep attachment to and love of the land, who in future would have no role to play in the growing of the food the country needs. Nor is it to argue that the past humiliation of the oppressed can only be assuaged by the future humiliation of the oppressors. One of the main functions of a new constitution would be to guarantee conditions in which all citizens, independently of race, colour, or creed could make their contribution to society and live in dignity and peace. But it is to insist that there be no *de facto* constitutional freezing of the present unjust and racially enforced distribution of land. There might be good arguments for the careful study of transition arrangements, for giving the present owners alternatives to sabotage and fighting to the death, for taking care to maintain high levels of food production while new generations of agricultural scientists are being trained, and for creating the conditions in which a common patriotism involving all South Africans is allowed to evolve. But these are essentially pragmatic factors that belong to the political arena. They are not inalienable human rights principles that belong to a Bill of Rights.

From a human rights point of view, the starting point of constitutional affirmation in a post-apartheid democratic South Africa is that the country belongs to all who live in it, not just to a small racial minority. If the development of human rights is the criterion, there must be a constitutional requirement that the land be redistributed in a fair and just way, not a requirement that says there can be no redistribution except on conditions that are clearly unattainable by the black majority.

Misconceptions about structure and implementation — the question of affirmative action

Since most proponents of a Bill of Rights in South Africa see it as an instrument designed to block rather than promote any significant social change, they completely omit from their projections any reference to affirmative action. This deprives the Bill of Rights of its true potential as a major instrument of ensuring a rapid, orderly, and irreversible elimination of the great inequalities and injustices left behind by apartheid.

Without a constitutionally structured programme of deep and extensive affirmative action, a Bill of Rights in South Africa is meaningless. Affirmative action by its nature involves the disturbance of inherited rights. It is redistributory rather than conservative in character. It is not a brake on change but rather a regulator of change, designed on the one hand to guarantee that change takes place, and on the other hand that it proceeds in an orderly way according to established criteria. Affirmative action enables all the interested parties to make an appropriate contribution, or at least to know where they stand. It presupposes the concertation of diverse forces in an agreed direction, with the State playing an ultimately decisive, though not necessarily exclusive role in the process. A Bill of Rights cannot accordingly be seen in the eighteenth century way simply as a fetter on the state in relation to the citizen (though it should never lose its character as a guarantee against abuse of citizens' rights by the state). On the contrary, through giving constitutional backing to affirmative action, it gives to the state, as well as other forces, a duty to use national institutions and resources to promote the rights of citizens.

In the historical conditions of South Africa, affirmative action is not merely the corrector of certain perceived structural injustices. It becomes the major instrument in the transitional period after a democratic government has been installed, for converting a racist oppressive society into a democratic and just one. It is the instrument in terms of which agreed national and constitutionally established goals are realized in a fundamental way, attributing appropriate responsibilities to all social forces — the public sector, the private sector, and the individual citizen.

Misconceptions amongst the mass of the people about a Bill of Rights

The way in which a Bill of Rights has been projected in South Africa as a means of preserving vested interests and of blocking affirmative action to bring about genuine equality has given the whole concept a bad name amongst the mass of people. This is most unfortunate. A Bill of Rights as such is neither a reactionary nor a progressive document. Everything depends on the context.

The fact is that there is a true and progressive concept of a Bill of Rights that merits the support rather than the suspicion of all genuine anti-apartheid fighters. This progressive concept situates such a document in its classic context as a true consolidator of the gains of people in struggle. Those of us engaged in the anti-apartheid fight also have our decisions to make. Either we leave the question of a Bill of Rights to others and then criticize the results, or we enter the fray directly and say 'these are our positions, this is where we stand, this is what a Bill of Rights should really be like'. More concretely, we can transfer the debate from the remote arenas of the think tanks and locate it where it belongs — in the midst of the life and strivings of the people. Justice and human rights do matter to us. This is what we are fighting for and there should be no cynicism in our hearts on the matter.

In South Africa we already in fact have a document that embodies the key elements of a Bill of Rights. It is a document that was born out of struggle, responds directly to South African conditions, expresses the aspirations of the oppressed people, and meets with internationally accepted criteria of a human rights programme — namely, the Freedom Charter adopted at the Congress of the People in 1955.

From a human rights point of view, the Freedom Charter was amongst the most advanced documents of its time. In clear and coherent language, it defends fundamental legal, political, and civil rights, it spells out social and economic rights that were only to become internationally agreed upon in the 1960s, and refers to people's rights that were only to be formulated in the 1970s and 1980s. The Freedom Charter is accordingly a contribution towards world human rights literature of which we South Africans can be proud.

A Bill of Rights for a post-apartheid South Africa: some pre-conditions

A Bill of Rights can either be an enduring product of history shaped by lawyers, or a transitory product of lawyers imposed upon history. If in South Africa it is to be the former and not the latter there will have to be, it is suggested, four basic pre-conditions:

- ☐ An appropriate process whereby a Bill of Rights may be adopted.
- ☐ In its content, the Bill of Rights must be associated with the extension rather than the restriction of democracy.
- ☐ The Bill of Rights must be centred around affirmative action.
- ☐ The mechanisms for applying the Bill of Rights must be broadly based, and not restricted to a small class of judges defending the interests of a small part of the population.

An appropriate process whereby a Bill of Rights may be adopted

Bills of Rights can be either copied, defined, negotiated, or constructed.

Copying a Bill of Rights

The easiest and least rewarding procedure is simply to copy a Bill of Rights from a model regarded as working well in another country. Apart from the fact that this saves on legal drafting fees, there is little that can be said in its favour. An effective Bill of Rights in any country must relate to the culture, traditions, and institutions of that country, at the historic moment when the Bill of Rights is considered necessary. This is not to deny an educative and exemplary role for a Bill of Rights, nor to refuse it a capacity to take on new meanings in the course of time. But it is to insist that an effective Bill comes from inside the historical process, not outside, and that it reflects a set of values gained in the course of struggle and rooted in the consciousness of the people, not one imported from other contexts.

Defining a Bill of Rights

Defining a Bill of Rights has the advantage of being directed towards the specific problems of a specific situation. This is what the burgeoning think tank movement on southern Africa aims to do: select experts who define their way into the problem and define their way out again. The flaw of this approach is that it presupposes that the basic issue is an intellectual one: if only the correct formula can be found, everyone will come to their senses, apartheid will disappear, and all will end well. The fact is that the basic problems are ones of power and consciousness, not of formulation. It is not chauvinistic to assert that there is no lack of brains in South Africa. Even the defenders of apartheid, unfortunately, have an intelligentsia of considerable brainpower, today armed with all the intellectual apparatus of what is called contemporary political science. The fact is that until the social reality and especially the power structure has changed, the intellectual reality will remain imprisoned. The context will be that of re-arrangement rather than substitution. Yet, try as the think-tankers might, there is no way in which apartheid can be adapted or modified to make it consistent with any meaningful Bill of Rights. Similarly, there is no way in which a Bill of Rights that obeys international standards could be adapted to be consistent with apartheid, however rearranged. Any constitutional scheme designed to entrench the rights of the white minority, whether property rights or rights to racially exclusive education or residential ares, violates the principles of equal dignity and equal opportunity which lies at the heart of any Bill of Rights. Unfortunately, most of the think tanks seem to have set themselves just such an agenda, namely, to propose a constitutional scheme which, under the guise of a Bill of Rights, would guarantee that however many blacks would be in Parliament, none of the privileges presently enjoyed by the whites would be touched.

Negotiating a Bill of Rights

Negotiating a Bill of Rights, the third method, has two great virtues. It operates from inside the process and, by definition, its outcome will correspond to the power realities of the moment, giving it a fair chance of becoming operational. But it has certain serious drawbacks. As in the case of a copied or defined Bill of Rights, the people who are to be the holders of the rights are regarded as mere spectators in the process. Furthermore, the negotiations inevitably result in a document so full of compromises and short-life arrangements that it hardly constitutes a true Bill of Rights. The fact is that one cannot negotiate goals. One has to establish them. What one can negotiate is the means whereby agreed goals can be achieved. If there is no agreement on goals, save at the level of banalities (such as that everyone shall be happy and none shall feel oppressed), then there is no basis for negotiating a Bill of Rights. In the case of South Africa, it is only when the fundamental goal of a non-racial, democratic, and united South Africa is accepted, that a suitable foundation could exist for negotiating the terms of a Bill of Rights. What could be negotiated then would be the precise configurations, both substantive and institutional, as well as the exact steps to be taken to get there in as speedy and orderly way as possible.

Even granted agreement on goals, however, the major weakness remaining is the passivity of the people at large in relation to their fundamental rights. We live in an age in which every form of communication with and involvement of the people is possible. Even in the difficult circumstances of apartheid South Africa in the 1950s, the meetings that preceded the Congress of the People at which the Freedom Charter was adopted involved hundreds of thousands and possibly millions of people. All the participants felt thereafter that in some way or another the document was theirs, made by them, for them, and for all the people of South Africa, something for which people were willing to fight, and, as Nelson Mandela said, something for which if needs be, they were willing to die.

Negotiations have an important role to play in removing obstacles to popular participation in the processes leading up to the final drafting of a Bill of Rights but they are no substitute for democratic input.

Constructing a Bill of Rights

In my view, this is what Bills of Rights are, or should really be about. This is also what makes the fourth procedure for adopting a Bill of Rights for South Africa imperative — namely, constructing such a Bill. A constructed Bill of Rights will, of course, copy from other models. It should eventually be a coherent and well-defined document drawing on the advice and experience of all the thinkers — whether in think tanks or outside — or the world. It will also involve important elements of negotiation. But in addition it will have the characteristics of:

- being built up over a period of time rather than drafted at one moment;
- being constructed in sections and layers rather than as a single, unique document; and
- being the product of active involvement of the widest strata of the population at all important times.

These three characteristics are interrelated. The time-frame gives the people as a whole and all special interest groups a chance to be involved. A Bill of Rights is built up, stage by stage, starting from agreement on general principles, and moving to specific institutional arrangements. In the meanwhile, all the major social forces that accept the basic goals are specially, though not exclusively, involved in the evolution of sections that have particular relevance for them. Thus, we already have in South Africa an education charter in draft, emerging in the course of struggle against racist education. One could contemplate a workers' charter in which trade unions would have a special role. Another example is a charter on religious freedom and the role of the churches, mosques, synagogues, and temples in promoting the goals of the new society. The embryos of important sections of a future Bill of Rights are thus already emerging in the work of the National Education Crisis Campaign, the workers' charters, the declarations of activist religious leaders, programmes of the women's organizations, and so on. At a future stage, when a democratic government has been formed or is imminent, the process of consultation and involvement could be extended and formalized. The Freedom Charter itself is, of course, an important document already in existence. On its foundation, a Bill of Rights could be gradually constructed, drawing upon all the inputs of all the different sectors.

In the same way as a constructed Bill of Rights presupposes a building up of the substantive part of the Bill, so it takes account of the need to involve, step by step, the institutions which are to be invoked to make the Bill operative.

Clearly it would be presumptuous to attempt to lay down or even predict the exact course whereby future constitutional documents will be adopted. But the perspective that needs at least to be considered is that of constitution-making as a process rather than an event. Once this is done, the possibilities become greatly enlarged of involving the people directly in the shaping and formulation of the rights of which they are to be holders. Rights in the true sense of the word are never conferred — they are seized, shaped, expressed, and lived by their bearers. In this way, the social contract ceases to be an abstraction and becomes a reality. The sovereignty of the people takes on a real and not purely notional meaning. Negotiating the terms is the final touch, not the foundation of the new constitution.

In terms of its content, the Bill of Rights must be associated with the extension rather than the restriction of democracy in South Africa

To project a Bill of Rights as being essentially a mechanism to frustrate majority rule is to doom it from the start. Of course, in entrenching certain fundamental freedoms and liberties, it establishes a rigid framework of rights within which majority rule will operate. These basic rights can be defended in the courts even against the will of the majority, yet they do not operate to deprive the majority of the right to deal with substantive questions facing the nation. The fundamental argument of this paper is that a Bill of Rights should precisely be used to enlarge rather than freeze the area of human rights, and to eliminate rather than perpetuate racial distinctions and the fruits of discrimination.

What needs to be done is to turn the Bill of Rights concept from that of a negative, blocking instrument, which would have the effect of perpetuating the divisions and inequalities of apartheid society, into that of a positive, creative mechanism, which would encourage orderly, progressive, and rapid change.

At the level of content, this would take into account specific features of the South African situation. There are a large number of areas that are relatively non-controversial and on which agreement could rapidly be reached. These include general civil, political, and legal rights. Yet property rights are highly contentious in the context of the impact of apartheid.

In relation to second generation socio-economic rights, attention needs to be given to breaking out of the confines of the Anglo-Saxon legal tradition whereby rights are basically restricted to what is justiciable, that is, to interests that can be protected by direct recourse to a court of law. While the role of the courts should always be important, it should be complemented by a richer concept in terms of which the Bill of Rights not only operates to defend individual rights, but seeks to guarantee the extension of rights to the community as a whole. To take one example, what would be more important: the right to sue your doctor or the right to health? Appropriate enforcement mechanisms should be created, such as sanitation control, safety measures, inspection, a system of primary health care, and school-feeding, all with appropriate legal underpinning.

Consideration thus needs to be given to a Bill of Rights as a programme and not simply as a set of directly justiciable interests. A constitutional document that is partly programmatic in character presupposes that certain major social goals are set out in the document. Public and private entities are placed under a legal duty to work toward their realization. The second generation of rights lend themselves more to treatment of this kind than to the justiciable first generation kind.

Third generation rights, such as the right to peace, development, and a clean environment also necessarily have a strong programmatic character which might be upsetting to lawyers habituated to Anglo-American legal conventions. The argument that such concepts are political rather than legal makes increasingly less sense in relation to the changes required in a post-apartheid society. The law has to recognize its political functions and politics have to be structured by law. The object is not to separate the two, but to find the right relation between them. International conventions and domestic legislation all have their role to play in defending these rights.

The Bill of Rights must be centred around affirmative action

The third fundamental feature of a meaningful Bill of Rights for South Africa is that it be structured around a programme of affirmative action. It is not just individuals who will be looking to the Bill of Rights as a means of enlarging their freedom and improving the quality of their lives, but whole communities, especially those whose rights have been systematically and relentlessly denied by the apartheid system. If a Bill of Rights is seen as a truly creative document that requires and facilitates the achievement of the rights so long denied to the great majority of people, it must have an appropriate affirmative strategy. To adapt Anatole France, if the law in its majesty were to give equal protection to a family of ten living in a two-roomed shanty and a family of two in a ten-roomed mansion, it would not be enlarging the area of human freedom in South Africa. Whatever attitude is taken to unused or under-used accommodation, the failure to impose a legal duty on the state and the private sector to reduce inequality in living conditions would be to deprive the Bill of true meaning in at least one important area.

The advantage of affirmative action is that clear and irreversible goals with an undeniable social and moral purpose are stated. However, considerable flexibility is permitted in terms of how the goals are to be realized. This helps avoid the dangers of backsliding on the one hand, and producing grandiose but highly voluntaristic and unrealizable plans on the other. If the private sector is to play a positive role in reducing inequality in a democratic South Africa, it is difficult to see how any strategy other than that of massive affirmative action could function.

The example of housing has been given. But just as there is no area of South African life that apartheid has left untouched, so it will be necessary to extend affirmative action to every aspect of South African society — health, education, work, leisure, to mention but a few. The transformations will have to affect not just the social and economic life of the population, but the very institutions of government. Even with the best will in the world, structures themselves built on inequality and injustice cannot be expected to be the guardians of justice and equality for others. In the presence of one of the worst wills in the world, the need to apply affirmative action to the civil service and the organs of state power becomes even more urgent.

The mechanisms for applying the Bill of Rights must be broadly based, and not restricted to a small class of judges defending the interests of a small part of the population

If the objective is to guarantee the minimum disturbance of existing property and social 'rights' (one has to put the word in inverted commas — the power to ensure that your child goes to a whites-only school cannot be dignified by the word 'rights'), then who better to fulfil the role than those who not only belong to and share the values of the very group to be protected, but whose whole professional mode has been shaped in the context of the interests and values of that group? If, on the other hand, the dog is to watch the interests of the formerly oppressed, it would have to have a totally different pedigree and training. The question of whether the word 'and' in a particular context only means 'and' or can also mean 'or', which has exercised the minds of lawyers for generations, would have little interest for defenders of the rights of the oppressed, who would look overwhelmingly to social rather than semantic factors in making their decisions.

This raises the important and delicate question of the relationship of a Bill of Rights to the legislative power of Parliament. It has already been argued that the objective of a Bill of Rights should be to reinforce rather than restrict democracy. In South African conditions, it is unthinkable that the power to direct the process of affirmative action should be left to those who are basically hostile to it. In later years, when the foundations of a stable new nation have been laid and when its institutions have gained habitual acceptance, it may be possible to conceive of a new-phase Bill of Rights interpreted and applied by a 'mountain-top' judiciary. At present, the great need is to give people confidence in Parliament and representative institutions, to make them feel that their vote really counts and that Parliamentary democracy serves their interests.

The kind of body that might provide a bridge between popular sovereignty on the one hand, and the application of highly qualified professional and technical criteria on the other, would be one similar to the Public Service Commission. A carefully chosen Public Service Commission with a wide brief, highly technical competence, and general answerability to Parliament, could well be the body to supervise affirmative action in the public service itself. Similarly, a Social and Economic Rights Commission could supervise the application of affirmative action to areas of social and economic life, while a Land Commission could deal with the question of access to land. Finally, an Army and Security Commission could ensure that the army, police force, and prison service were rapidly transformed so as to make them democratic in composition and functioning, perhaps the hardest and most necessary of all the tasks facing those who wish to end apartheid in South Africa. At the same time, the courts would have a fundamental role, far more important than their role today. Representative of the people as a whole, they would be the bastions of first generation rights, which would be fully justiciable, and the guarantors by means of judicial review that second and third generation rights were realized in a constitutional way.

Summary

To sum up: the oppressed and all true democrats in South Africa have a great interest in promoting a Bill of Rights for the country and seeing it as a welcome and progressive phenomenon. But such a Bill of Rights has to be created over a period of time with the active involvement of the people. It has to be located in the heart of the democratic process and not be seen as a foreign object imposed upon it. It also must be structured around a strategy of affirmative action. Finally, its implementation has to be entrusted to institutions that are democratic in their composition, functioning, and perspective, and that operate in a manifestly fair way under the supervision of the people's representatives in Parliament and subject to review, in terms of the constitution, by the courts.

Such a Bill of Rights, born out of the struggle for freedom, would live for decades, perhaps centuries, and enrich the international patrimony of human rights.

The question of majorities and minorities

Apartheid has the capacity of turning the banal into the marvellous. The principle of equal rights, which in most other countries is regarded as so ordinary as not to merit any explanation or require any defence, is projected as something quite wondrous in South Africa, indeed so astonishing as to be constitutionally illusory and practically unattainable.

The anti-apartheid struggle is directed precisely towards the achievement of full equality between all South Africans, independent of race, colour, ethnic origin, gender, or creed. The measure of the success of any new constitutional order will thus be the degree to which it enshrines the principle of full, genuine, and ineradicable equal rights.

Equal rights mean equal rights for each and every individual South African. As far as the basics of citizenship are concerned there will be no distinction whatsoever between persons on the grounds of race or ethnicity. Just as race classification and group areas will disappear from legislation, so will they vanish from citizenship and the electoral system. There will be a common voters' roll, made up of and speaking in the name of the whole nation. In this sense, the constitution will be completely colour-blind and totally race-free. There will be no special privileges for racial or ethnic groups, no vetoes, no areas of special competence, or 'own affairs.' Race will only enter the constitution as a negative principle, that is, to the extent that the constitution is not only non-racial but anti-racist. The anti-racist character will be guaranteed by provisions, expressly referring to race, which:

☐ outlaw racial discrimination,

☐ prevent the dissemination of racist hostility, and

☐ ensure that measures are taken to overcome the effects of past racial discrimination.

The question is raised as to what guarantees would exist in such a constitutional order, especially one based on majority rule, against persecution of minorities by the majority. It may be said that, even recognizing the evils of apartheid, it would be unjust to inflict on future generations of whites the very kinds of discrimination that their fathers and mothers have been and are inflicting on blacks. At the more pragmatic level it may be contended that if one wishes to persuade whites to relinquish power now, they must be given reasonable guarantees against persecution in the future.

Like so many questions in South Africa, the issue of minority rights is presented in a back-to-front way. International human rights law pays considerable attention to the issue of minority rights. In fact, protection of minority rights preceded and opened the way for international protection of individual rights. Yet always the context was the protection of minorities against discrimination and persecution. The object was never to accord to minorities the right to discriminate against and oppress others, as happens in South Africa today. Nor has the protection of minorities ever encompassed the maintenance of special privileges for a minority who constitute a social and economic élite. The whole object of minority rights law has been to counteract discrimination and inequality, not to perpetuate it.

More recently, protection of minority rights has moved beyond the principle of non-discrimination and equal rights, and added the principle of non-assimilation, that is, of maintaining the right to cultural, linguistic, or religious identity in the face of pressure to adopt the ways of the majority. In other words, the law has both affirmed the right of minorities to be the same as the majority in terms of basic civil, legal, and political rights, and different in respect of language, cultural, and religious rights. In fact, international law increasingly accepts the right to use affirmative action or positive discrimination procedures to promote the language, culture, religious, and even economic rights of minorities that have been subjected to past discrimination.

The complication in South Africa is that minority rights law is being invoked as a basis for preserving the interests of a dominant and not a dominated minority. It is paradoxical that in South Africa it is the majority and not the minority that would be the natural subject and automatic beneficiary of minority rights law. It is the majority that has been discriminated against and kept out of public life, and it is the majority whose languages and culture have been despised or else presented in a distorted way. If affirmative action has any scope, it is to overcome the effects of discrimination against the majority, not the minority.

What all sections of the population in South Africa have a right to expect — and this would include the group self-classified as whites — is that any future constitutional set-up protects them against discrimination and abuse, and recognizes their right to preserve and develop their cultural, linguistic, and spiritual heritage. If this were all that the white minority were after, then a new constitution would have been drafted long ago. Just as certain fundamental rights and freedoms can be guaranteed for the individual citizen as part of the context in which majority rule operates, so certain rights for minorities can be guaranteed as constituting the overall legal framework within which majority rule takes place. It would place certain parameters on the scope of majority rule, without undermining the principle of

Parliament functioning according to the principles of majority rule. In fact, the Freedom Charter contemplated exactly such a relationship between majority rule and respect for cultural and other rights when it declared that the people shall govern, and that all national groups shall have equal rights.

Unfortunately, what really has been in issue is the perpetuation of constitutional privileges under the guise of establishing legitimate defences against domination of group over group. Hence the recourse to such bizarre constitutional devices as separate voters' rolls, own affairs, and group vetoes, all of which have a profoundly racist character, and the use of such offensive phrases as the need for protection against being swamped.

In fact there are quite simple ways in which the principle of non-domination can be maintained without recourse to racist concepts and structures. Three levels of constitutional devices can be contemplated, each preventing arbitrary or unjust treatment or harassment of any kind on the basis of race, appearance, origin, religion, or language. These devices supplement the general rights of citizens, complement each other, and in their conjunction acknowledge the cultural richness and diversity of the country. They could also be associated with a non-racial electoral system that guarantees political pluralism and provides protections against political oppression.

In the first place there should be a Bill of Rights which entrenches basic individual rights for citizens. Any individual discriminated against on the grounds of belonging to any minority (or majority), that is, on the grounds of race, colour, ethnicity, language, or gender, would have appropriate legal recourse. This is the *guarantee of equal individual rights*.

Secondly, there should be a general non-discrimination provision which will outlaw any discrimination against any group as a group on the grounds of race, colour, ethnicity, language, or religion. Any member of a group discriminated against would have legal remedy even if he or she is not directly affected by the discrimination. This is the *guarantee against discrimination*.

Thirdly, the cultural diversity of the country would get a strong degree of constitutional recognition so as to permit groups to develop certain aspects of what they might call their own way of life with a view to enriching society as a whole. This is the *guarantee of equal rights for all national groups*. Here it is necessary to separate out from a group's 'way of life' what are presently objectionable features that require abolition, what are really universally or widely accepted modes of behaviour not restricted to that group, and what are intimate characteristics that justify protection and even promotion.

The right to behave as a member of a master race, to insult blacks, and to use violence gratuitously, for example, are boasted about by some as a marked feature of the way of life of a certain group today. Clearly these would be denounced in any new democratic constitution.

Similarly, there are many social habits which in reality belong to, or are open to all people, such as matters of dress, cuisine, and etiquette. One does not need a constitutional right to eat curry or have a braaivleis (barbecue) or wear trousers. What will be guaranteed will be the inviolability of the home, freedom to pursue family life, and general freedom of personality. None of these freedoms attach to a particular racial or ethnic group.

Next, there are certain activities that historically and culturally have become associated with certain groups, usually based on linguistic association. Thus there are historically created communities with a distinctive socio-cultural personality possessing considerable subjective significance for its members, and contributing towards the general overall texture of South African life. Shorn of their association with oppressive domination, these socio-cultural features will continue and even have a measure of constitutional protection and support. What will not be permitted is basing political rights on socio-cultural formations, nor attempts to restore apartheid by political mobilization based on setting group against group. Thus, from a general juridical and citizenship point of view, the whites as whites will disappear from South Africa, as will the blacks. There will only be South Africans. But within the framework of an equal and undivided citizenship, there will be full recognition of linguistic diversity. That is, there will be one South African citizenship with a single suffrage, but many South African languages.

There will be extensive recognition of the right to constitute religious organizations, many of which may have their holy literature in a particular language. Afrikaans-speakers who feel comfortable worshipping in the Dutch Reformed Church will be able to continue their prayers and hymns in the way to which they are accustomed, as well as to choose their spiritual leaders, and to develop their doctrine according to the internal teachings of the Church. In this sense there will be unfettered freedom of religious-cultural association (one can think of many other groups — Jews, Muslims, Hindus, Greek Orthodox, as well as the many African independent sects that might have a similar basis). What would not be permitted would be to deny membership on grounds of race etc. Nor would these socio-religious organizations be allowed to function as a cover for political mobilization on a divisive, racist, or ethnic basis. One hopes, in fact, that the religious organizations will play an active role in helping to build a united South Africa and to overcome the inequalities and divisions left behind by apartheid. Without their involvement, the task will be difficult indeed.

Another sector where the constitution could manifest a special tolerance could be in relation to certain areas of traditional law and custom. This is a question where extensive discussion with the people is required. All that is rich and meaningful to people can be retained and progressively developed; while that which is divisive, exploitative, and out of keeping with the times — especially that which has been distorted by colonialism and apartheid — can be eliminated.

Finally, it should be mentioned that there will be other constitutionally-protected group rights that by their nature will necessarily cut across linguistic and ethnic divides. Thus the workers of South Africa, who today are playing a key role in the fight to destroy apartheid and build a new South Africa, will receive extensive constitutional recognition in the form of both individual and collective rights. Similarly, South African women, also in active combat, who have been the victims of special social and legal disabilities, should have the right not only to be free from discrimination but to call upon special resources so as to overcome the legacy of past discrimination. Other groups that could merit special constitutional recognition might be children, the aged, handicapped persons, and victims of apartheid persecution. In none of these cases would the question of race or ethnicity enter.

The Afrikaner businessman and the African peasant

The above considerations could perhaps be best understood if applied to a concrete situation. For the purpose of making a clear projection into the future, imagine how the adoption of a democratic constitution could effect two prototypical persons — an Afrikaner businessman and an African peasant — and then see what significance the constitution would have for the relationship between the two. Simply to say that both will enjoy equal rights is not enough. At present, their relationship is one of profound inequality. The question arises as to how the constitution would promote real, not simply formal, equality between them. Furthermore, it is necessary to reflect on the cultural-linguistic dimension, which, while disappearing as a basis for the exercise of political rights, nevertheless continues to be relevant in relation to cultural and national rights.

Looking first at the position of the Afrikaner businessman in relation to the new constitutional order we see that:

☐ As a citizen he will enjoy all the civil and political rights which he presently exercises in his privileged capacity as a member of the dominant racial minority, but will do so on the basis of being an equal citizen, exercising normal constitutional rights in a non-racial, democratic South Africa. Thus, he will have the right to elect and be elected, to join the political party of his choice, and to criticize or defend the government. He will also have the right not to be deprived of his liberty except in terms of the law and after a fair trial. He will enjoy freedom of speech and information, but will not continue to have the right to propagate division and hatred on grounds of race.

☐ With regard to personal rights, he will have security in his home, the right to live a family life if he so chooses, the right to enjoy his hobbies and pastimes, the right to move freely around the country, the right to have his holidays, and the right to visit other countries.

☐ As a businessman he will continue to have the right to exercise his professional and entrepreneurial skills and to be appropriately remunerated therefor. His rights to personal property (a home, a motor car, a bank deposit, etc.) will be protected, while his rights to productive property will be recognized but subject to the principle of public interest and affirmative action.

☐ As an Afrikaner, he will have a guaranteed right to use and develop his language and to belong to the Dutch Reformed Church (non-segregated) or to any other religious body of his choice. If he wishes as part of his private life to mix with and marry only Afrikaners, that will be his choice. Similarly, there will be no interference with the habits and customs of daily life, most of which will in fact be practised by many non-Afrikaners. What he will have to learn to live with, however, is that in relation to anything outside the immediate private or family sphere, there will be constitutional norms of non-discrimination. Thus there will certainly continue to be schools and universities in which Afrikaans is the medium of instruction and in which special attention is given to the study, development, and enrichment of the Afrikaans language and literature. But these schools would not be able to restrict their entry on the basis of race.

☐ Similarly, Afrikaners might continue to occupy certain neighbourhoods as a matter of social practice. What they would not be able to do would be to create racially exclusive areas to which non-Afrikaners or non-whites were not admitted. The new constitution thus would not only automatically declare void and repeal such divisive legislation as the Group Areas Act, but would also prohibit the use of restrictive covenants or the formation of racially exclusive condominiums as a means of continuing apartheid, this time in a privatized form.

Looking next at the position of the African peasant, we find that:

☐ For the first time he will be able to enjoy full and normal rights of citizenship, especially those of suffrage, in the land of his birth. He will no longer be subject to arbitrary arrest, removal, or banishment. All the apartheid laws which presently dominate his life will be annulled.

☐ As a person, he will for the first time be free to move and reside anywhere in the country. His home will be inviolate. His dignity as a person and his right to a stable family life will receive full constitutional protection.

☐ As a farmer he will have a claim on the state for access to land and to technical, educational, and financial support. As a property owner, what possessions he has will be protected. His house will be safe from the bulldozers; his plot of land and livestock guaranteed against confiscation. He will have a claim on the state to assist him to build, buy, or rent a decent home and to enable him to acquire an interest in land for farming that will be legally protected.

☐ As an African he will for the first time enjoy equal rights with all his fellow South Africans and be free from any discrimination or deprivation. His language will be recognized as one of the national languages of the country. His culture and the history of his forebears will also be respected. Place names, national monuments, and national holidays will record the struggle of his and previous generations. As a victim of past discrimination and domination, he will have a claim on the state for invoking the procedures of affirmative, corrective action.

The above analyses have proceeded on the basis that the personalities are male. If they are female, an extra constitutional element will enter, namely the Equal Rights clause, which will bar any discrimination on the grounds of gender. In addition, women will have an affirmative action claim to remove disabilities or disadvantages associated with past discrimination. Women will also have constitutionally recognized benefits in relation to maternity and to child care.

Carrying the constitutional projection one step forward, and positing that the African peasant is a tenant farmer on land owned by the Afrikaner businessman, what bearing would the future constitution, and especially the Bill of Rights, have on their relationship? In broad terms, the constitution will require that the immense injustice, whereby eighty-seven per cent of the land belongs to a fifteen per cent minority, be corrected as rapidly as possible. Exactly how this will be achieved and how this will affect the specific relationship between the businessman and the farmer, will be conditioned by two factors, one historical and the other institutional.

The historical factor relates principally to the behaviour of the businessman. If he and his class prefer to fight to the death, if they threaten to destroy and massacre the workers as a protest against the installation of a democratic government, then they should not be surprised if appropriate countermeasures, including confiscation of land and goods, are taken. If on the other hand they accept a new patriotism, adhere to the new constitution, and continue to use their productive skills for the growing of food and for the benefit of the country as a whole, the process of land redistribution will necessarily have a less drastic character. Affirmative action presupposes orderly, significant, and irreversible progress to eliminate the inequalities produced by centuries of colonialism and apartheid. As has been stated, constitutionally determined criteria must establish clear goals. Then the parties most directly interested must negotiate the means whereby these goals can effectively be achieved. If disputes arise on the modalities of change, appropriate conflict resolution machinery exists, with the courts playing a key role in ensuring that correct procedures have been adopted and appropriate criteria applied.

In the case of land, it is of course not the soil itself that is redistributed, but title to or interests in it. Here the possible legal forms are infinite, ranging from state confiscation, to outright state purchase, to joint ventures with the state (or local public authorities), to co-operatives, to non-racial private or public companies or corporations, to partnerships, to parcelling off land to individual farmers. Regional particularities and the existence or otherwise of abandoned or unused land will be relevant, as will, to a considerable extent, the economic importance to the country of maintaining high levels of food production. Similarly, the time needed for new owners, shareholders, partners, or co-operative members to acquire appropriate technical and management skills will become a relevant consideration; it could be five years, or ten, or fifteen; it could be after death. Legally enforceable, phased arrangements could be worked out. The particular wish and family situations of the interested persons and their historical relationship to the land could also be taken into account, as well as indebtedness to the Land Bank and inheritance taxes.

What is certain is that the present deformed and unjust relationship between the Afrikaner businessman and the African farm-tenant, structured on legally protected arrogance and domination, will come to an end. Equal citizenship will not just be a formal measure. It will have concrete content, enabling every person whatever his or her background, to act as a free person and to enjoy the benefits of freedom in the land of his or her birth.

Conclusion: transitional arrangements

The only value of predictions about the future is to enable their makers to determine at a later date how far from the mark their original prognostications were. In the case of South Africa, a tense battle is underway assuming many forms, and although the eventual defeat of the forces of apartheid can be predicted with certainty, the precise time that this will take and the nature of the intermediate or transitional phases are still open.

Thus, if apartheid is eventually destroyed by insurrection and a revolutionary seizure of power, the correlation of forces will be such that the classes of society represented by the victorious revolutionaries can impose their terms on society as a whole. A constitution is necessary to institutionalize the new power, not to bring it into being. It will include a Bill of Rights, but the procedures of affirmative action to ensure the restoration of land, wealth, and dignity to the people would inevitably be far less cumbersome and protracted than those contemplated here.

On the other hand, the prospects of a negotiated settlement have improved considerably, even though there are great pitfalls ahead. The position of the anti-apartheid forces has long been that the making of a constitution for a democratic South Africa belongs to the people as a whole, acting through a democratically elected Constituent or National Assembly. What should be negotiated is not a constitution, but the transitional arrangements leading up to the making of a constitution and, possibly, the procedures to be followed and the basic principles to be applied by the Assembly. By their nature, such arrangements — which might or might not include political and legal guarantees of a firm though transitory kind — will fall short of the democratic ideal. For their reduced life span, they could well include certain features that still bear lingering imprints of apartheid society. Such transitional arrangements must, however, be distinguished from the failed attempts to create a so-called internal settlement like the Zimbabwe-Rhodesia set-up of Bishop Muzorewa and the Turnhalle Agreement in Namibia. In the first place, such internal settlements were arrived at by means of an alliance between the apartheid rulers and a black collaborator class. Since the majority of the people were excluded from the agreements, nothing was settled, the war continued, and the only difference was that blacks played a bigger role in the oppression of their fellow blacks. Internal settlements are meant to be permanent, whereas transitional arrangements are intended to be self-eliminating; in short, internal settlements are a means of postponing democracy, while transitional arrangements are a means of hastening it.

The negotiation of transitional arrangements could in fact pave the way for a relatively peaceful dismantling of the structures of apartheid and the establishment of a democratic South African state. The goal of a race-free society would not be negotiable, but the means of getting there, and in particular, the time-table and method of transferring power from a racial minority to the people as a whole, would be.

In this context, it becomes more important than ever that opponents of apartheid the world over keep their eyes fixed on the goal of genuine democracy in South Africa. To suspend pressure because apartheid managed to don attractive new clothes or shed some of its dirty old apparel would be to betray generations of South Africans who have struggled to free their country from racial oppression and exploitation. It would also be to postpone the peace which we all so sorely need, and delay the reconstruction necessary to ensure that South Africa truly becomes a country that belongs to all who live in it and a proud member of the community of nations. The reward for ending apartheid is democracy. When democracy comes, it will no longer be necessary for anyone to fight against the isolation of South Africa; sanctions and the boycott will fall away automatically — indeed, the world will embrace the new nation with special warmth.

2 Evolving a Bill of Rights culture

Very few South Africans know what a Bill of Rights is. It is something outside our experience, something we associate with seemingly exotic legal and political cultures like those of the USA. The battle for human rights in our country has essentially been a struggle for the vote and not for a Bill of Rights.

We do not know what it means to have a constitutionally entrenched Bill of Rights. We are unfamiliar with the notion of constitutionalism and constitutional rights. We are still betrothed to the idea that rights come from law-makers, not that the law enshrines our rights. Yet when people marched in the streets without asking for permission, they were asserting the fundamental right of peaceful assembly. When they defied the bans and restrictions imposed under emergency regulations, they were affirming that as South African citizens they had basic rights, whatever the authorities might have said. When workers went on strike illegally, they were not merely campaigning for better conditions, but expressing their right to withhold their labour, irrespective of potential legal penalties. The foundations of the constitutional idea already exist in the day-to-day behaviour of millions of people. Just as racism and authoritarianism go together, so do democracy and constitutionalism. Indeed, one sometimes fears that authoritarianism is even more deeply ingrained in what is called the 'South African way of life' than is racism. People can hardly decide anything for themselves, do anything on their own — they always have officials over them taking decisions on their behalf. Sometimes the rulers are kind, sometimes they are cruel. But always it is they who decide. Constitutionalism creates a totally different relationship between citizens and government. Not only does it make government accountable to the people rather than people to the government, it also establishes the idea of people having rights and of people being equal. The constitution is there for everybody, and everybody should be able to appeal to it at all times. The anti-apartheid idea achieves its highest legal expression in the form of a new constitution. This will be *our* constitution, created by South Africans for South Africans, a product of struggle and thought. It is up to us to ensure that the constitution embodies the principles that we have been fighting for, that it materializes and secures the transition from the anti-apartheid phase to the pro-democracy one.

All modern constitutions establish the framework of government and indicate how the government is to be selected, whether by election, hereditary succession, and so on. Most have a Bill of Rights as well, setting out the fundamental rights and freedoms of citizens. Sometimes the Bill of Rights is justiciable, that is, it contains mechanisms for its enforcement through the courts. In other cases it merely proclaims rights which citizens ought to have, without specifying implementation mechanisms. Clearly, the new South African constitution will require a justiciable Bill of Rights which will incorporate the fundamental rights and freedoms as set out in the Universal Declaration of Human Rights and the Freedom Charter.

There are two common misconceptions about the role of a Bill of Rights. One is that it should attempt to resolve in advance all the fundamental political problems of the country. This is not its function. The Bill of Rights creates the framework within which people freely debate the great national issues and choose their government. It is certainly not the function of a Bill of Rights to foreclose public discussion and choice in relation to major social and economic issues. That is what elections and Parliament are for. It is, thus, important to distinguish constitutional from electoral issues. Equal protection of the law and the redress of structural inequalities resulting from past discrimination are constitutional, not electoral questions, whereas the debate on regulation or deregulation, privatization or nationalization, are electoral and not constitutional ones. Similarly, the right to health, education, and housing can be set down as constitutional objectives without specifying in advance how they should best be legislated for.

The second misconception relates to the connection, frequently expressed in negative terms, between a Bill of Rights and majority rule. It is often said that a Bill of Rights is needed in South Africa as a counter to majority rule. This is an incorrect and misleading representation of the relationship between a Bill of Rights and majority rule. The fact is that in a democracy, far from negating majority rule, a Bill of Rights guarantees that the true voice of the majority is heard. If there is no freedom of speech and assembly, if there are no free and fair elections, if there is no way of getting rid of an unpopular government, then there is no means of guaranteeing that rule is really by the majority and not by a self-perpetuating minority. Similarly, the entrenchment of the basic rights of due legal process and inviolability of the person and the home, are not restrictions on majority rule but part of the democratic and law-governed framework within which majority rule functions. Majority rule says that the people shall govern, not that the people shall oppress. The exclusion of tyranny from the purview of majority rule frees it rather than freezes it. What is kept at bay is not the tyranny of the majority or tyranny of the minority but any kind of tyranny. Once disassociated from any link with abuse and arbitrariness, majority rule comes into its own and guarantees the full ebb and flow of democratic opinion.

Apartheid denies both majority rule and a Bill of Rights. We do not even speak about basic constitutional rights; in fact, we have never had a constitution in the sense of a fundamental, entrenched law, which cannot be easily amended and which provides the framework for the adoption of other laws. With the exception of the two entrenched clauses referred to below, the Acts of Parliament called constitutions have been no more than ordinary statutes subject to amendment in the same way as any other statute. In legal terms, they have no more weight than an Act providing for a fixed ice-cream price.

When the Union of South Africa was created in 1910, Parliament was given unlimited sovereignty, subject only to certain vestigial links to the British Empire which were gradually erased. The Union of South Africa Act united what had formally been four British colonies into a single state, and laid down the structure of government and procedures whereby the legislature would be elected, the executive established, and the judiciary nominated. In the British parliamentary tradition, nothing was said about the rights of citizens. It was a laconic constitution, proudly technical and devoid of programmatic or human rights provisions.

Two provisions of the Act were entrenched, namely, the right of a certain number of blacks in the Cape to remain on the voters' roll, and equality between English and Dutch (later Afrikaans) as the official languages. However, these provisions were not protected by a Bill of Rights; the entrenchment mechanism merely required a special parliamentary majority for change. Attempts by the National Party in the 1950s to by-pass this special majority led to the Appellate Division of the Supreme Court nullifying a series of Acts of Parliament. These cases turned on the required parliamentary majority, not on questions of fundamental human rights.

The only judge who ever attempted to exercise a testing power and declare legislation to be unconstitutional and therefore null and void was Kotze of the Supreme Court of the Boer Republic of the Transvaal. Modelling himself on Marshall and other great American judges, he tried to strike down certain Acts of the Volksraad because they did not comply with the Constitution of the Republic. Paul Kruger, the Boer President, simply ignored him, in fact, dismissed him. For various reasons, neither the British nor the Boers subsequently claimed Kotze, and his actions have been regarded by scholars as curiously aberrational rather than as the beginnings of a Bill of Rights tradition.

The absence of a Bill of Rights tradition, however, has not implied the non-existence of a human rights tradition. Even if we do not include primary resistance as part of the human rights tradition (when people defended their lands and independence with spear in hand), we had anti-slavery agitation and a struggle for a free press as far back as the 1920s, the movement for African rights which began emerging in the 1880s, the campaigns over the treatment of Boer women and children in concentration camps at the turn of the century, and the feminist movement shortly thereafter. We have also had the struggle for the Afrikaans language, passive resistance campaigns throughout much of this century, trade union struggles, the adoption of the Freedom Charter in 1955, and the whole contemporary anti-apartheid movement. There are many personalities of whom the current generation of human rights activists can be proud — Pringle, Rubusana, Gandhi, Abdurahman, Schreiner, Seme, Plaatje, Junod, Krause, Gumede, Luthuli, Fischer, Gqabi, First, and others.

Similarly, at the purely technical level, and even without express constitutional authority, South African courts have created for themselves a certain amount of space within which to exercise a moderate degree of judicial review. They have not been able to refer to any clear constitutional base for this power, using common law principles and English judicial precedent instead.

Thus, judges have from time to time upheld claims that proclamations, regulations, and by-laws issued by bodies or persons in terms of powers granted by Parliament, be declared void because of vagueness or unreasonableness. Judges have also invalidated certain executive acts, such as forced removals or banning orders, because persons adversely affected were not given a hearing.

At times these judges have softened the impact of apartheid legislation by applying what are legally referred to as statutory presumptions. Statutory presumptions could be invoked if there were gaps or uncertainties in a statute, enabling the courts to favour the interpretation least onerous to those whose rights were adversely affected. This was argued on the basis that, as Parliament itself was a product of liberty, it could be *presumed* to have intended a pro-liberty construction on its legislation. (However, if legislation was clear and unambiguous, the judges gave full effect to it, even if it violated fundamental human rights.) The word 'presumed' has to be emphasized. The judges attributed to Parliament a will in favour of liberty and fair dealing that was truly fictitious. They declared that unless the enabling Act took away rights by clear language or necessary implication, such rights should be presumed to exist; for example, the right to be heard before being made to suffer a penalty or disadvantage, or the right to be clearly informed as to what behaviour constituted a crime, or the right of access to one's lawyer.

The reality, however, was that Parliament, at its next session, almost always refuted the generous assumptions imputed to it, preferring to regard any presumed pro-liberty intention as an irritating loophole to be plugged. Parliamentary draftsmen were then called upon to be more astute in future and to eliminate any possible inference of legislative respect for individual rights.

The adoption of a Bill of Rights would accordingly effect a considerable change in the relationship between the judiciary and Parliament. Judges would not merely be interpreters of laws adopted from time to time, they would be defenders of the constitution, with power to decide whether any government activity or law was constitutional or not. For this reason, a new constitution must be carefully drafted to ensure that the rights of all are protected and apartheid eliminated as speedily as possible. In addition, the judiciary should enjoy the respect of all South Africans, with the backing of a constitutional court representing the wisdom of all sections of the population.

To the extent that Bills of Rights have appeared on the legal scene, they have done so in the most negative context possible — as provisions or declarations in the documents referred to as the bantustan constitutions. This has been doubly unfortunate. Bills of Rights have been linked with manoeuvres to defend apartheid by imposing a spurious independence on the so-called ethnic homelands. Whatever the intention of the authors might have been, such Bills of Rights have done little, if anything, in practice to mitigate harsh arbitrary rule in the bantustans. Human rights in these areas have been advanced by popular struggles rather than by the courts.

On the principle of good coming out of bad, however, we can benefit from the bantustan Bill of Rights experience by learning negative lessons. Thus we can assert the following:

☐ In order to be meaningful, a Bill of Rights must be associated with democracy, not a verbal patina covering an authoritarian regime.

☐ The people affected by the Bill of Rights must be involved in the process of its formulation, so that they regard it as their own, something for which they have struggled and something they will defend, even if in a particular case its immediate application is inconvenient to many of them.

☐ The content of the Bill of Rights must correspond with the deepest aspirations of the people, and have a manifestly just quality.

☐ The people as a whole must have confidence in those who guard over the Bill of Rights, and they must see themselves and their highest virtues reflected on the Bench.

Close encounters of an intellectual kind

Two recent events have completely transformed the nature of the discussion on a Bill of Rights for South Africa. The first was an ANC statement early in 1986 supporting a justiciable Bill of Rights to protect the fundamental rights and liberties of all individuals in the country. This was followed by the publication of the organization's Constitutional Guidelines, which defined how the Bill of Rights would fit into the total constitutional picture, and more particularly, how it would relate to programmes of affirmative action.

More recently, the South African Law Commission, instructed by the government to inquire into the viability of a Bill of Rights for South Africa (and especially to see how group rights could be incorporated in such a document), issued its report. Among many debatable points, it declared firmly that a Bill of Rights would be meaningless without the basic right to vote, and that a Bill of Rights should not be designed to protect the rights of racial or ethnic groups but of individual citizens.

Thus, two independent processes of inquiry, undertaken against completely different experiences of life, referring to quite separate sources, and using totally distinct modes of discourse, found themselves reaching similar conclusions, or rather, recommending similar points of departure. The gap is still large, but they are talking about the same thing.

Such close intellectual encounters are so rare in our divided country that when they do occur the possibilities they offer should be fully explored. The fact is that a wide democratic and anti-apartheid consensus is beginning to emerge in South Africa, representing the convergence of forces and personalities that previously had little to do with one another. People from the Mass Democratic Movement are speaking to liberal democrats and to Afrikaner democrats, each drawing on different traditions and life experiences. We have all travelled a long way to get there. The growing consensus between them provides a solid basis for fruitful discussion by a wide range of lawyers concerned with the question of human rights in our country.

Frequently, lawyers find themselves out of touch with reality because their proposals are idealized projections into the future. The danger now is the opposite: that reality will leave us behind. What we articulate are not just private musings of legal dreamers, but the wishes and longings of millions of our compatriots, a clear majority in our land, drawn from every nook and cranny of the country, and representing all its diverse social formations.

The struggle for human rights takes place everywhere at all times. It reaches directly into all structures of our society, so that every individual, even those involved in institutions which, until now, have been used for purposes of oppression, has an opportunity to make her or his contribution.

Clearly it is necessary to consolidate the broad anti-apartheid consensus and reinforce the sense of shared goals. We should determine precisely what the areas of common thinking are and then embark upon a constructive dialogue on our differences. In some cases we shall convince each other, in others we shall be able to compromise because the issues are relatively tangential, while there will undoubtedly be various questions on which we simply have different opinions. We need to find out how we can handle these disagreements: should we leave them to the democratic process; should we subsume them in a general compact which gives people most of what they want without fulfilling anyone's desires completely; should we try to agree on transitional arrangements whose objectionable features are attenuated by the fact that they are declaredly short-lived; or should we simply leave the matter over to be solved by future generations? There are many possibilities.

What matters is that we are learning how to agree and how to disagree. There is an important relationship between the two processes — it is precisely because we are agreed that apartheid is an abomination and a disaster that we are able to come together and discuss our disagreements on how best to achieve its abolition. It is the existence of the consensus on the one hand that permits disagreement on the other. Indeed, the right to debate questions freely and openly, a right consistently denied to the majority since the early days of conquest, is one of the most fundamental rights for which we are fighting. We cannot wait until Day One of the post-apartheid society to begin debating our differences and finding ways of handling divergent points of view. We need to acquire the habit now.

Thus, perhaps even more important than knowing how to agree, is knowing how to disagree. By debating our differences openly and on a basis of equality, we exemplify our aspirations.

In broad terms, there is wide agreement on the need to end apartheid in South Africa and to establish a democratic society, as generally understood throughout the world. Such a society should recognize the equal worth and dignity of every citizen, and provide appropriate protection for his or her fundamental rights. More specifically, we recognize that a future constitution should contain provisions which establish fundamental rights and freedoms. Such basic liberties have to be acknowledged and respected by the legislature and the executive, however sizeable the majority may be at any moment in favour of ignoring them.

Furthermore, these constitutional provisions should not be merely aspirational, but capable of speedy invocation through clearly identifiable and secure mechanisms. More concretely, citizens should have the right of recourse to an independent judiciary, respected by the population at large and heeded by whatever government is in power at any time. In a phrase, we favour a parliamentary democracy subject to a Bill of Rights.

We are aware that in Britain a major debate is taking place over the desirability or otherwise of adopting a Bill of Rights. Although we follow this discussion with interest, we note that in our country the theme of the protection of human rights has a special dimension which makes even those who would otherwise opt for unqualified majority rule favour the adoption of a Bill of Rights.

Strong and clear protection of individual rights on a non-racial basis makes the protection of group rights on a racial basis not only objectionable but unnecessary. A Bill of Rights coupled with guarantees of an orderly transition to full democracy provides far more security to all South Africans than racially based constitutional schemes which entrench the racial principle as the dominant feature of public life and ensure that voters are forever mobilized on racial grounds.

Once white South Africans can accept the simple fact that they are just people like everyone else, and not the lords and mistresses of anyone, they will enjoy far more security under a Bill of Rights than they would in the precarious constitutional laager of group rights.

There is nothing to stop us taking a look at human rights documents in other countries and on other continents. We all participate in an international human rights culture and share in the patrimony of human rights instruments. To the extent that we have succeeded in showing that apartheid is an international issue, so are we correspondingly obliged to heed international expectations with regard to apartheid's elimination.

What we look for abroad are human rights documents, not human rights consciousness; the formulations can be imported, not the awareness. Indeed, the frequent and massive human rights violations in our country, together with a vigorous movement of contestation and considerable international attention, have produced on our part unusual sensitivity to and a passionate interest in the safeguarding of human rights.

For those of us who have suffered arbitrary detention, torture, and solitary confinement, who have seen our homes crushed by bulldozers, who have been moved from pillar to post at the whim of officials, who have been victims of assassination attempts and state-condoned thuggery, who have lived for years as rightless people under states of emergency, in prison, in exile, outlaws because we fought for liberty, the theme of human rights is central to our existence. The last thing any of us desires is to see a new form of arbitrary and dictatorial rule replacing the old.

Yet, confident as we are in the strength and resilience of our South African-born human rights convictions, we can only benefit from the great store of human rights wisdom accumulated in many other countries over many centuries. This is particularly true in relation to how best to organize institutions to guarantee respect for fundamental human rights.

Indeed, while forever insisting on the specificity of our experience and our solutions, we firmly deny any idea of South African exceptionalism. The universalization of the human rights idea is one of the great achievements of our era. What we want is for this universal idea to link up with our own special strivings and become a living force in our land. We want simple justice, simple democracy, simple freedom, no more no less.

Thus we have no difficulty in agreeing on the following universally recognized fundamental principles:

- the equal dignity and worth of all South Africans, irrespective of race, colour, gender, origin, or creed;
- the inviolability of the person, the home, and correspondence;
- freedom of movement, residence, and travel;
- the right to vote, stand for election, and engage in political campaigns;
- freedom of expression and the right to information;
- freedom of conscience and the right to practise one's religious faith;
- the outlawing of torture and of cruel, inhuman, or degrading treatment;
- the prohibition of servitude, slavery, or forced labour; and
- the requirement of due process of law in relation to any deprivation of liberty or imposition of penalties.

By their nature, human rights documents know no copyrights; indeed, there is a certain resonance, a certain sense of security, to be gained from utilizing tried and tested formulations.

We live in an era of internationally held views on what constitutes the minimum guarantees of a fair trial. There are many precise formulations describing the circumstances that permit deviations from the basic protections accorded to citizens — such as requirements of public health or natural and other disasters. We may profitably look to the formulations adopted in the major international human rights conventions and charters, especially those of the United Nations, of Europe, the Americas, and Africa.

Once it is accepted that there should be a Bill of Rights and that all the classic, universally agreed-upon rights should be included, the really interesting part of the debate begins.

In addition to serving the goals of any Bill of Rights anywhere in the world, a South African Bill of Rights should address the specific problems raised by the fact that we will be moving from an apartheid to a post-apartheid society. Different sections of the population are likely to look to a Bill of Rights for different things. Many will hope it protects them against the kind of abuse to which apartheid has subjected them over the decades and guarantees to reduce and eliminate the enormous inequalities and indignities under which they are living. Others will see in it a bulwark against destructive revenge and a guarantee against the collapse of the economy and the disintegration of all social norms. The big problem will be how to integrate the basic longings and fundamental expectations of all South Africans, however diverse, in a single document; moreover, to do so in a way that ensures that the constitution is both operative and manifestly fair. The ideal Bill of Rights should, in addition to its classic functions of protecting basic civil, political, and legal rights:

☐ help remedy and eliminate the injustices, indignities, and inequalities produced by apartheid;

☐ create a climate of tranquility conducive to a good quality of life and to economic advancement; and

☐ promote the building of a nation.

3 To believe or not to believe

Nkosi Sikelel' i Afrika — God watch over Africa ... this is the popular anthem of national liberation and of peace, sung throughout southern Africa by believers and non-believers alike. The words are religious, the occasions secular. For most of its life, the ANC, a secular organization, had a national chaplain, one of whose functions was to lead the annual conference in prayer. There is strong opinion in favour of reviving the office today.

We want a secular state in South Africa, but a secular state with religion, indeed, with many religions. Religion is there, it is part not only of the spiritual life of people, but of the country's culture, of the very culture of struggle. Strip South Africa of its multitude of religions and it is no longer South Africa. Take away the anthem, Regina Mundi church, and St George's Cathedral, the Tutus, Naudes, Boesaks, Mkatshwas, Solomons, Chikanes, and internationally the Huddlestones, and the anti-apartheid struggle is no longer the anti-apartheid struggle.

To believe or not to believe — that is one of the most important constitutional questions facing any country, and one of the most significant philosophical issues facing any individual. It is a question which each one of us answers in his or her own way. In some respects it touches on what might be the most fundamental human right of all, certainly the most intimate and personal, the right to conscience. No one should be compelled by the state or by anybody to believe, nor should anyone be forced not to believe. Belief by its nature is something personal and intrinsic to the individual. It belongs to the conscience of each one of us, but also has a social dimension, a cultural dimension, even a national dimension.

The state should be neither theocratic nor atheist, but secular, tolerant, and accepting of the deep importance religion has for millions of South Africans. Religious communities, for their part, should be free to organize their worship as they please, and encouraged to take part in the life of the nation.

This is one of the most important areas for asserting the simultaneity of the right to be the same and the right to be different. In terms of general civil, political, and legal rights, all South Africans have the right to be the same, independently of their beliefs. One's rights as a voter or litigant or patient should not be affected by whether one is a Catholic or a member of the Dutch Reformed Church or a Methodist or an adherent of an independent African church or a Muslim or a Jew or a Hindu or an atheist.

The word creed used in the hallowed statement that we are fighting for a South Africa where all enjoy equal rights, irrespective of race, colour, or creed, refers exactly to this. Respect for the rights of conscience and acknowledgement of the basic worth of each individual citizen, require explicit constitutional recognition of the right to be the same, independently of one's belief or non-belief. The state neither rewards nor penalizes the presence or absence of any particular belief. The applicable constitutional principles are those of non-discrimination, equal protection, and personal privacy, that is, of the rights of conscience in a defensive sense.

At the same time, we have the right to be different. This involves the right to believe or not to believe, the right to worship in our own ways, to organize our own religious communities, to consecrate our own holy places, and acknowledge our own holy texts, to appoint our own religious leaders, and follow our own rituals and dietary practices. At the constitutional level, this raises questions of the right to religious expression, freedom of association, and the rights of privacy or personal conscience in an affirmative sense.

If there are Muslims who wish to attend mosque on Fridays, and Jews who refuse to ride on a bus on Saturdays, or Christians who believe it is sinful to catch fish on Sundays, that is their business and their right. They cannot impose their beliefs on others, nor can anyone take their beliefs from them. These are not rules that the state should impose in a secular, multi-faith society, but rules that believers voluntarily adhere to in their capacity as members of a spiritual confession. The sanctions for disobedience lie in the spiritual rather than the secular sphere, save that the religious community may require penance or decide upon expulsion. The state does not impose penalties.

Thus, most countries in the world permit lotteries, leaving it to the individual to decide whether taking a lottery ticket is simply a remote way of fulfilling financial dreams or an impious interference with the Lord's right to control chance. Similarly, in nearly every part of the world, Sunday is regarded as a day of recreation and entertainment as well as a day of worship and rest. The constitutionality of laws prescribing the values of some believers for the whole society would have to be examined in the light of the principle of the right to be different. Religious observance is not an area for the imposition of the sameness of duties. Sunday observance should be a matter of personal conscience in the context of the norms of a religious community, not a question for state law.

These are issues which should be worked out as far as possible by the religious organizations themselves. The state for its part should adopt a respectful and sensitive attitude in response. There would be many areas requiring collaboration. To be secular does not mean to be anti-religious, but rather that there is no official religion, no favouring of any particular denomination, and no persecution of or discrimination against non-believers.

The general principle of separation of church and state could be followed (the word church including mosque, synagogue, and temple). Yet it need not be absolute. The state could continue to maintain legislation recognizing the authority of certain religious leaders to register marriages as marriage officers. It could encourage non-proselytizing persons to offer religious comfort in hospitals and also respect religiously-based dietary rules in hospitals and other state institutions. It could co-operate with schools, hospitals, and other social institutions run largely or exclusively by religious organizations. It could discuss with religious leaders the circumstances in public life where prayers would be said or oaths taken. It could make appropriate arrangements for the broadcasting of religious services. It could permit non-denominational places of prayer at state institutions, including in the buildings of Parliament itself. It could adopt *Nkosi Sikelele* as the national anthem.

The problem is how to find a coherent constitutional format to take account of this active relationship. Both the state and religious bodies are anxious to preserve their respective areas of sovereignty, yet both wish to collaborate in a harmonious way in relation to areas of common concern. There is a wide range of constitutional options available for governing the relations between religious organizations and the state. They are:

- ☐ Theocracy, that is, the acknowledgement of religious organizations as the holders of public power and of religious law as the law of the state.

- ☐ A partly secular, partly religious state, with legal power-sharing between the state and religious institutions — each exercising constitutionally recognized power in its own sphere, usually with religious bodies controlling family law and, possibly, criminal law, and the state controlling all other aspects.

- ☐ A secular state with active interaction between the state and religious organizations, which not only have a constitutionally recognized sphere of autonomy, but collaborate with the state in tasks of mutual concern.

- ☐ A secular state in which religious organizations have a tolerated, private sphere of action, but there is no overlapping or joint activity with the state.

- ☐ A secular state in which religious organizations are repressed.

These are questions which should be worked out in a mutually acceptable way by all interested parties. Clearly there is no scope at all for the suppression of religion, nor is there any possibility of having a state religion in South Africa, nor of giving religious organizations judicial or other authority beyond the voluntary authority accepted by members. It would seem that in the light of South Africa's history and culture, something along the lines of the third option mentioned above would achieve the greatest support, namely, a secular state with active interaction between the state and religious organizations.

Ideally in South Africa, all religious organizations and persons concerned with the study of religion would get together and draft a charter of religious rights and responsibilities. This would be a direct contribution towards enriching the texture of a new constitution. It would not exclude them from commenting on all other aspects of the constitution, but would guarantee their voice being heard directly in relation to issues of the most immediate concern to them.

The constitution would in its Bill of Rights set out the broad principles governing freedom of religion, while the charter would elaborate more concrete principles and procedures. The charter would be attached to the constitution and have legal status as an entrenched and not easily amendable part of the law of the land; it would deal with the legal rights of religious bodies and individuals, not with ecclesiastical matters in themselves. The charter could, in the first place, establish constitutional norms guaranteeing the fundamental right to conscience.

We may be living in the most materialistic of worlds, (not of dialectical materialism, but of market materialism); freedom of consumer choice may be represented by some as the foundation of all freedoms, so that in their view the fundamental constitutional principle should be one person, one calculator.

Yet life surely will be more than one gigantic non-racial supermarket, and the quality of existence in the new South Africa will hopefully not be measured simply by degrees of material gratification. Conscience can make heroes of us all. The basic constitutional tenet in a free South Africa must be the right to think what we want to think, and to believe what we want to believe — one person, one conscience. The inviolability of the individual, the respect for the worth of every human being, whether seen as God-given or as natural endowment or as social right, lies at the foundation of citizenship. It was this right more than any other that apartheid violated.

The charter of religious rights and responsibilities would naturally deal with the constitutional protections required to enable religious organizations and communities to function freely. These could be spelt out broadly, or itemized specifically. They could cover everything from the inviolability of places of worship to respect for burial sites to recognition of the full right of religious bodies to interpret their doctrine and choose their leaders. They could include rights to remain in contact with spiritual brothers and sisters abroad, to go on pilgrimages, to publish holy documents and commentaries, and to establish seminaries and training schools.

It would be up to the participants themselves to define what they consider to be their fundamental rights. Certain groups may have very special concerns. There are, for example, millions of adherents of so-called independent African churches who may wish to spell out provisions guaranteeing them constitutional recognition without requiring them to surrender their autonomy. The implications of special beliefs such as those held by Jehovah's Witnesses in relation to blood transfusions or military service, could be resolved in as non-conflictual a way as possible.

Indeed, the whole question of conscientious objection in a democratic South Africa would require sensitive handling. Pacifists and pacifism have a major role to play in the new, democratic South Africa. Once apartheid is abolished and steps embarked upon to eliminate the inequalities created by apartheid, the major objection currently being raised to performing military service would fall away.

The army could possibly become an important agency for promoting cultural interchange and a spirit of genuine, non-racial patriotism. Yet there will continue to be persons who on religious and philosophical grounds refuse to bear arms. Many of them will in fact have played an active and courageous role in the struggle against apartheid. Appropriate legal arrangements should be made to acknowledge their rights of conscience in the form of alternative service of a constructive and non-humiliating kind.

This raises an even wider question: once apartheid is abolished, does the role of religious organizations in keeping witness come to an end? Clearly not. It is precisely their role in the anti-apartheid struggle today that gives them the moral authority to bear witness in post-apartheid society.

One does not have to believe in the Devil to accept that evil is multifariously seductive. Society as a whole can only benefit if, basing themselves on principles of fundamental morality and respect for the dignity of every human being, religious organizations in a democratic South Africa make their contribution towards maintaining respect for basic human rights and fair dealing.

Further, one hopes that religious bodies will not simply be passive critics on the margins of the processes of eliminating inequality and building a new nation, but active participants at the centre of things. Together with the trade unions, the religious organizations constitute important nation-building agencies. They reach into every area of society and bring people together who otherwise might live completely separate lives. Clearly, their tasks are fundamentally spiritual, but the spiritual encompasses the process of healing, especially of healing the nation and healing the lives of individuals damaged by apartheid; it means encouraging tolerance, respect, and love, which in South African circumstances promotes a shared sense of common humanity and of being South African.

Religious organizations constitute important elements of civil society. People should be free to join and leave them, and to participate in their manifold activities, whether baptisms, marriages, or funeral services, or choir competitions, or cake sales.

Not all the activities of religious organizations in the past have been beneficial to South Africa. At different stages, religious bodies have justified every form of dispossession and humiliation. They have preached submission in the face of injustice, and blessed the rifles and machine-guns of the dispossessors. They have practised apartheid within their own ranks, and engaged in unseemly proselytizing competition with each other.

Different religious fundamentalisms in their more aggressive forms have promoted disunity. Among the polychrome perils alleged to have faced South Africa — the red variety, the black, and the yellow — have been the diverse dangers of Rome and of Islam, not to speak of constant conspiracies of the Elders of Zion or of international atheism. Satan seems to have taken time off from all his or her other worldly duties to perplex the people of South Africa.

Yet, other traditions have never been lost — the insistence on an ethical basis for personal conduct, the spirit of service and community, the idea of universal brother- and sisterhood, the dedication to good deeds. Generations have been inspired by the vision of a world in which swords are beaten into ploughshares, in which neighbours love each other as themselves, in which there is no distinction between gentile and Jew, in which the poor are blessed and the meek inherit the earth. The poetical and mystical visions of the Holy books have entered the world views of most South Africans and become a rich part of the national imagery and way of looking at the world.

These are powerful points of reference for the creation of a new united South Africa, in which national life is enriched by religious diversity, and religious organizations transform themselves and become both more spiritual and more South African as they help transform the country.

4 Free speech — unlimited or qualified?

South Africa has suffered so many interferences with the rights of free speech that the tendency to let people say what they want, when they want, how they want, is very strong. At the same time there is an awareness that racism can ignite explosive passions and destroy the very fabric of a tolerant and democratic society. Furthermore, it is impossible to gloss over the fact that, in addition to being unjust and exploitative, apartheid is spiritually injurious, insulting, and defamatory. The problem, then, is how to reconcile the need for openness and the right to speak one's mind with the necessity for healing the wounds created by racism.

Clearly the constitution must protect the normal rights to criticize the government and public officials, to take part in free public debate over issues confronting the country, to discuss international questions. People should have an unqualified right to argue for or against socialism or capitalism or abortion or capital punishment, or to warn us that the end of the world is near. Similarly, if the Flat Earth Society wishes to establish a branch in our country, they should be free to do so — there will be no lack of potential adherents.

Yet the real problem is not tolerance to the flat-earthers or more seriously, to the nationalize-everything-ers or the privatize-all-ers. Nor is it whether or not to have free speech corners where every Tom, Dick, or Harriet can mount his or her soapbox. The real issue is what to do about the organized mobilization of racial and ethnic hatred.

Many countries have legislation which outlaws group libel. Should the South African constitution permit and even protect the right to say such insulting and provocative things as that all whites are rapists who should be driven into the sea? Or that blacks are baboons who should never have been given the vote? Or that the Xhosas have come to

Natal to suck the blood of the Zulus? Or that the Shangaans are cowards and never knew how to fight? Or that South Africans of Indian origin should be deported to India? Or that Hitler knew how to treat the Jews?

In South African conditions, these are fighting words, the language of pogroms and blood. There is a strong argument for saying that if the constitution is a compact, agreed upon by representatives of all the major groups in South Africa, it should include a shared undertaking not to indulge in mutual insults and not to permit the mobilization of rabid racist or ethnic feelings for political advantage. In this sense, democracy and non-racism become inseparable — there is no democratic right to be racist. You do not have to love your neighbour, but you can be prevented from insulting him or her.

In theory, the constitution can adopt one of three positions in relation to racist speech: it can protect it, it can leave the question entirely to the legislature, or it can lay down express qualifications in relation to free speech, including prohibition of defined forms of incitement to hatred and division. If it adopts the third position, the further question arises of how best to combat the promotion of racial hostility — whether to rely on the criminal law or civil restraints or voluntary codes of conduct affecting the media and political organizations, or whether to include provisions in the electoral law which forbid the creation of parties on racist principles or campaigning on the basis of racist or tribalist emotion.

There are other questions which bear indirectly but significantly on the question of free speech, and which could affect the way constitutional principles are formulated. At present the press in South Africa is anything but open and anything but non-racial. The *Rand Daily Mail*, the most informative and widely-respected daily paper of the 1960s and 1970s, was closed not on journalistic grounds, but because its circulation was too high among blacks who had no money and not high enough among whites who had money. In absolute market terms, nothing should be free, not even speech.

English-language and Afrikaans-language monopolies control virtually the whole of the commercial press, which means virtually the whole of the press, and not only the press itself, but most of printing and distribution. Similarly, broadcasting is almost entirely in the hands of the state.

What the commercial and state monopolies have in common is that they are completely white-dominated, locked into the apartheid structures. This affects not only the appointment of journalists, but the very determination of what is front-page news.

Some attempts have been made by generations of courageous and imaginative journalists, both black and white, to mitigate the effects of this inequality. Space has been won for black voices in the commercial press, while journals such as *New Nation*, the *Weekly Mail*, and the newly-established *Daily Mail*, have transformed reporting in South Africa. There is also a large number of dynamic community-based alternative media, and highly intelligent critical journals.

Yet huge obstacles exist to the free flow of information in South Africa, ranging from unequal degrees of literacy to the underprivileging of many languages to official secrecy to conscious or unconscious biases in the presentation of news. The new oral tradition of resistance has proved far more resilient and informative to the mass of the population than has the media. Yet we cannot rely on oral tradition in the new democratic South Africa to keep people informed.

At the same time, we must remember that the objective is to open doors that are presently closed, not to create more blockages to the free circulation of ideas and information. We would have gained little if we were to replace the present media controls with new ones that simply switch the propaganda and biases around; if one realm of banality takes over from another. Truth has always favoured the democratic cause, and our people are tired of forever being protected in the name of what others think is good for them.

As Thabo Mbeki, member of the National Executive of the ANC, has pointed out, freedom of expression will have a special significance in a new South Africa. People's expectations will be enormous, the means available for satisfying them scant. The new government can either shut people up and decide on their behalf what to do with the limited resources, or else involve people themselves in making informed choices. Clearly, the latter requires the maximum circulation of information and ideas. Freedom of expression and accountability thus become inseparable.

All these are issues which impinge on the language and substance of the constitution. We look to our articulate, technically experienced, and battle-scarred media people to lead the way in proposing solutions.

5 Judges and gender: The constitutional rights of women in a post-apartheid South Africa

Some call it the woman question. Some call it the man question. From a constitutional point of view it is best referred to as the gender question. Few would deny that gender is on the agenda, but not many would agree on how to formulate the question, and even fewer would lay claim to have found the constitutional answers. The issue is painful, discomfiting, and controversial, good reasons for tackling it.

It is a sad fact that one of the few profoundly non-racial institutions in South Africa is patriarchy. Amongst the multiple chauvinisms which abound in our country, the male version rears itself with special and equal vigour in all communities. Indeed, it is so firmly rooted that it is frequently given a cultural halo and identified with the customs and personality of different communities. Thus, to challenge patriarchy, to dispute the idea that men should be the dominant figures in the family and society, is to be seen not as fighting against male privilege but as attempting to destroy African tradition or subvert Afrikaner ideals or undermine civilized and decent British values. Men are exhorted to express their manhood as powerfully as possible, which some do by joining the police or the army or vigilante groups and seeing how many youths they can shoot, whip, teargas, club, or knife, or how many houses they can burn down or bulldoze, or how many people they can torture into helplessness. Patriarchy brutalizes men and neutralizes women — across the colour line.

At the same time, gender inequality takes on a specifically apartheid-related character; there is inequality within inequality, or, put another way, some are more unequal than others.

Any constitutional dispensation relating to gender must accordingly take account of both dimensions — the universal issues affecting women and men, and the specific forms that apartheid has given to gender domination in our country.

Thus, African women have pointed out that as a group they have suffered many layers of disability, some shared with other groups and some specific to them. They declare that they share with their African menfolk the experience of national oppression; with all South African women the burdens of inequality and sexism; with the workers of all races the problems of economic subordination.

At the same time, as African women, they are subjected to special disabilities and disadvantages. Accordingly, although they are oppressed as Africans, they are doubly oppressed as African women.

Colonialism and apartheid have progressively whittled away the democratic aspects of traditional African society and law and emphasized vertical power and patriarchy. The result has been to leave African women in limbo, stripped of the secure if junior position they had in traditional society and denied individual rights by state law. The space they occupy is that pitiless zone where the different patriarchies meet. For a century and more, traditional and state law have been interpreted in such a way as to relegate grown African women to the status of minors, subjecting them to the guardianship of fathers, brothers, uncles, and male cousins. This has most severely burdened widows, whose legal position in relation to the family home, holdings, and goods has been extremely precarious.

The whole issue of the future of African family law is essentially a cultural one which has to be treated with great sensitivity. Its solution will require extensive discussion, with primary involvement of those most likely to be affected by any change. The constitution could emphasize respect for tradition or it could underline the principle of equal rights. The two are not necessarily incompatible. Tradition could continue at the social level, enabling families to make such customary marriage arrangements as they wished. At the same time, the parties to the union would, as South African citizens, enjoy their constitutional rights, one of the most important of which would be equal treatment in law.

Thus there would be nothing to prevent the payment of lobola or *bohadi*, and the parties and their heirs could, if they chose, apply all the traditional rules that flowed from such arrangements. If the woman wished, as a matter of custom, to allow her husband to represent her in dealings with the outside world, that would be her choice and the law would not make such a situation illegal. Nevertheless, she would at all times be free to invoke her constitutional right to equality. If she felt that patriarchal rules were adversely affecting her rights to inheritance or a pension; or to custody of her children or maintenance for them; or her right to take up residence where she wished, or enter into a contract, or take up employment or travel; the law would come to her aid. (One assumes that the courts will be transformed so that the people will see themselves reflected on the bench both in terms of cultural background and gender; it would not be a case of one section of the community sitting in judgement on another, but of new family courts drawn from the whole population applying the constitutional principles in a firm but culturally sensitive way.)

Although they are oppressed as women, they are doubly oppressed as African women. A century of migrant labour and the pass laws has had a particularly injurious effect on the lives of African women, depriving them of sexual companionship, family life, and economic tranquility. One result of this is that the right to live a 'normal' life in the context of the nuclear family becomes a feminist demand in South Africa (just as in areas where polygamy still exists, monogamy enters the list of women's claims). The constitution would thus have to attempt simultaneously to protect the rights of unmarried women, single parents, widows, and divorcees, support the institution of the family, and protect women against the inequality created by patriarchy. This is no easy task, but certainly not impossible. The starting point must always be the claims and perceptions of the persons most affected, namely women.

Again, although they are oppressed as workers, they are doubly oppressed as women workers. Hundreds of thousands work as domestic servants, without trade union rights or legislative protection. They are frequently employed in the worst-paid jobs subject to the most inconvenient hours. Millions are involved in unpaid and backbreaking agricultural work, not to speak of the especially large amount of unpaid domestic work they put into their own households, having the least access to domestic help, labour-saving equipment, convenience food, and organized child care.

There are many other questions which bear most acutely on African women but which affect the lives of all South African women (and all South African men).

A multitude of issues exist in relation to gender and work, such as equal pay; discrimination in hiring, promotion and firing; the allocation of jobs on a gender basis; maternity and paternity leave; safety in relation to reproductive capacity; nursing rights and child care; working hours; sexual harassment.

Similarly, there are acute gender-related questions pertaining to health and control of one's body and reproductive capacity — issues ranging from the organization of health care delivery, to health education, contraception, and abortion.

Another set of questions relates to violence against women both physical and mental, direct and indirect. This includes rape and domestic violence, but also sexual harassment in its manifold forms, the demeaning use of women in advertising, and, many would argue, the degradation of women in pornography.

There are also sharp issues related to gender-biased use of language and gender stereotyping in public documents, educational material, the media, and advertising.

Finally, there are never-ending problems related to the family and family break-down, to the difficulties of single parents, to welfare support, and the rights of divorcees and widows.

Four different approaches have been adopted within the anti-apartheid movement towards the above issue clusters, each with profoundly different implications for any future constitution and Bill of Rights.

The first view, not as common now as it once was, is that to raise the gender question at this stage is to divide people when the goal around which we all should unite is the abolition of apartheid. Women will have the vote in a free South Africa and be able to vindicate their rights through Parliament in the ordinary way. Patriarchy is a vague concept, and in any event, it should be fought by means of education and not through the law.

The second position is that if a Bill of Rights is to be introduced at all, it will be meaningless if it ignores the rights of the female half of the population and permits discrimination against them on the basis of gender. The constitution must accordingly contain an unequivocal declaration in favour of equality between men and women, in terms of which all laws and practices which place either sex at a disadvantage shall be unconstitutional and void.

Thirdly, there is a growing body of opinion that formal equality is worth little if not supplemented by affirmative action to destroy the structures and behaviour patterns created by centuries of discrimination against women. Accordingly, special constitutionally-backed criteria and mechanisms are required to enable women to break through the layers of disability inherited from the past. Coupled with this approach is an insistence that instead of taking a completely gender-neutral approach, which in reality seeks to assimilate women into a world constructed around male interests and ways of seeing things, the constitution should permit and require the law to look at the actual lives that women lead and thereby enable women to define for themselves what their expectations and priorities are. It also requires a strong female presence and voice in all the processes leading to the adoption of a new constitution and attention to the language used in its formulation, so as to ensure that women feel the constitution speaks directly to and for them and does not simply tuck them away in some safe appendicular space.

Finally, there are those who argue that patriarchy and sexism are older and even more pernicious than apartheid, and that failure to construct a constitutional order expressly dedicated towards their abolition will result in the transition process from apartheid to a post-apartheid society being little more than the handing over of power from one gang of men to another.

Although each approach has considerable technical implications, the decision on which one to adopt is essentially political and not technical. Fundamentally, the question will be how strong the women's movement is, and how clear and united it is in its goals, which may be framed in ways quite different from those presented above.

The basic right underlying all other rights is the right of women to speak in their own voices, the right to determine their own priorities and strategies, and the right to make their concerns felt.

At the same time it must be said that the outcome of the debates in the women's movement is of deep interest to the whole of society. It will affect men and men's behaviour and assumptions, and it will have a decisive effect on the character of social transformation in South Africa.

Without the active involvement of millions of women, not simply as mobilized detachments for change, but as lively participants determining the very meaning and quality of such change, there can be little hope of achieving what may be regarded as the three goals of a post-apartheid constitution, namely, the eradication of the injustices of the past, the creation of a tranquil and prosperous society, and the building of a South African nation. It is not just that women constitute more than half the population: the social deformities and injustices created by apartheid fall with special severity on women, so that the rights of women and the ending of apartheid are inextricably linked.

Indeed, it is no longer fruitful to debate whether or not gender should be on the agenda. It is there already, having been put there by the women's movement in a way that cannot be ignored.

What follows is an attempt to anticipate the range of technical options available for achieving an integration of women's claims into the very substance of the constitution, and to look tentatively at some of the dilemmas involved. Once more it is necessary to state that the issues are highly complex and controversial, and that the processes whereby they are discussed are probably more important than any particular solutions proposed.

One of the first problems is whether the constitution and Bill of Rights should have any provisions dealing with women's rights or gender equality at all. This query is not as anti-feminist as it first appears. There are some who maintain that the presence of women should be felt in every article of the constitution, and not in any special provisions. Just as it is quite unnecessary for the constitution to outlaw discrimination between people who have blue eyes and those who have brown eyes, or between those born on a Monday and those born on a Tuesday, so should the accident of having male or female sexual organs be constitutionally irrelevant.

The rights of citizenship, they argue, should be as non-gender as they will be non-racial. If the constitution speaks of workers' rights and then goes on to refer in a different section to women's rights, it implies that women and workers are separate categories, whereas women constitute nearly half the paid work-force and ninety per cent of the unpaid.

For all the forcefulness of such arguments, it would seem that in reality constitutional silence would do nothing to abate gender-awareness and everything to permit the continuation of existing discrimination and abuse. What is valuable in the above approach, however, is that it requires that every clause of the constitution be infused with an awareness that it has been formulated by both women and men with a view to protecting the rights of both men and women. This means excluding the gratuitously obnoxious use of the word 'man' when persons of both sexes are meant (e.g. 'one man, one vote'). It also suggests that in order to underline the equality of citizenship and the inclusion of women at every stage, the word 'person' can, where appropriate, give way to the expression 'man and woman' or 'woman and man'. Thus a clause dealing with the vote could begin: 'every man and woman who has attained the age of ... ' while the preamble could repeatedly make it clear that the constitution was drafted by women and men for a citizenry composed of women and men.

Universalizing, emphasizing, and equalizing the presence of women and men in all situations might in fact do more to dissolve or de-construct imposed and artificial concepts of masculinity and femininity than would any attempts to ignore the question of gender.

Equal and explicit gender-presence throughout the constitutional text in no way impedes special provisions directed at eliminating gender inequality. On the contrary, it provides a secure foundation for an equal rights clause, emphasizing that it is not a case of women having the right to be equal to men, but of women and men being equal to each other.

Being equal does not mean being identical. The equal rights clause must be framed in such a way as to recognize the right to be the same in some areas, and the right to be different in others. Thus women and men are the same in their capacities as voters or litigants or office-holders or users of public facilities. They are different in terms of child-bearing (though not necessarily of child-rearing) and whether, historically and culturally, they have been the perpetrators or the targets of gender discrimination or abuse.

The constitution would thus not be violating the equal rights concept if it took account of this biological and socio-cultural reality. On the contrary, it is only by acknowledging this reality that the constitution can serve its true role of guaranteeing equal protection.

Similarly, equal rights provisions should not be phrased in such a way as to prevent affirmative action procedures in relation to overcoming the effects of past gender discrimination. Affirmative action does not require that unqualified women be given preference over qualified men, but it would permit special opportunities being created for women in the same general qualification bracket as men. More importantly, affirmative action would permit special programmes of education, training, and search in order to encourage women to qualify themselves for and obtain employment in areas to which they have previously had difficult access.

The equal rights clause should also not be formulated in such a manner as to imply that it and it alone covers the gender question. There are strong voices in the women's movement which argue that emphasis on equality can even obscure ways in which the law should intervene to correct the injustices to which women are subjected on a daily and massive basis. Women should be free to walk in parks and gardens and along streets; they should be safe from violence by husbands or lovers; they should not be subjected to being pawed or whistled at, or to seeing their bodies being used to sell commodities or degraded in sadistic pornography. For millions of women, the right to have a safe abortion could be more important than the right to enter medical school or become manager of a bank. The whole problem of sexism, of stereotyping, of the thousand little gender-based assumptions that make life stifling and oppressive, that take away confidence and deny any sense of completeness and fulfillment, is barely touched by an equal rights clause; as has been said, women's place is in the wrong.

These are issues of overwhelming importance to a very large section of the population. Men have their physical strength, their economic power, and the force of tradition behind them. In this context, they do not need the protection of the constitution, in fact do not even see that these are questions of a constitutional dimension. Surely, however, it should be possible to distill the preoccupations and demands of women into a constitutional clause that summarized their essence in terms of a basic human right and thereby provided the foundation of future legislative and judicial action to provide remedies.

One recognizes that questions which are essentially cultural and psychological in character and that touch on the most intimate aspects of human relationships cannot be resolved simply by legal declarations, even less by super police forces. Nevertheless, the constitution is a special document that should speak to the whole nation without fear or favour (except, possibly, favour to the oppressed). It establishes fundamental principles for the whole of society and serves as a point of reference for all. It inspires, it educates, and it creates institutions for implementation. One envisages, then, compact constitutional provisions which:

☐ proclaim equal rights and equal protection under the law;

☐ permit and require affirmative action to overcome the accumulated effects of discrimination; and

☐ seek to outlaw or at least move away from all forms of abuse, oppression, and insult based on gender.

Yet however vivid and complete these formulations may be, it is difficult to see them as being anywhere nearly satisfactory in themselves. There are important and complicated issues relating to the family, to health, to special problems of employment, to reproductive rights, to child care, to abortion, to unpaid work (to mention only a partial list), which do not easily subsume themselves under broad constitutional provisions.

In particular, it is necessary to reiterate that important sections of the women's movement have expressed their reservations about too much emphasis on equal rights and not enough stress on the right of women to affirm themselves as women and not as neutered men.

They point out that women live in their bodies, which are different from those of men, and grow up with a set of experiences of and perspectives on the world which differ from those of men. Their problems come from having to suppress their sensibilities and vision under the weight of constructs which pretend to be social and neutral but which really represent a male view of the world. They point out that in South Africa as in the rest of the world, men have not done very well with all their power, not only denying self-expression to women, but constantly killing, torturing, or locking each other up.

A distinction should thus be drawn between femininity, which is an idea that comes from men and is imposed on women, and feminism, which is a form of self-affirmation by women themselves. According to this approach, self-determination for women as a group and freedom for women as individuals, require a constitutional order which, far from suppressing or neutralizing women's distinctive voice, guarantees it space and opportunity to be heard. The constitution should therefore in this area permit what might seem the surprising doctrine of equal but separate; it would facilitate a right to be different that would cut across the gender-bar and give choice to men as well as women who felt oppressed by sexual stereotyping.

Another facet of this approach is the emphasis which it places on personal autonomy and the right to choose. Thus the key question should not be whether women as a group should be doing everything that men do — healing, killing, judging, or lorry-driving — but that each individual woman should have the right to choose for herself whether she wishes to be a professional or a trade union organizer or a welder or a housewife. The right of choice becomes especially important in relation to health questions and the issue of control of fertility. Thus the question of abortion could be dealt with on the basis of acknowledging the constitutional right of anti-abortion groups or individuals to campaign for women not to resort to pregnancy terminations, while at the same time recognizing that in the last resort it is the pregnant woman herself who has the constitutional right to make the final decision.

Indeed, the issues are so multi-faceted, intricate, and simultaneously concrete and elusive, that it is doubtful whether a few broad constitutional generalizations could in themselves ever provide sufficient guarantees — and one has to bear in mind that in the transitional period at least it will largely be male parliamentarians and judges who will be responsible for their interpretation.

The answer would seem to lie in the adoption of a charter of women's rights, formulated essentially but not exclusively by women and expressing their claims and priorities. Such a charter would aim to be declaratory, affirmative, educational, and operational, that is, it would declare the rights that women and men have, it would establish a programme of action to be undertaken to realize the rights in practice, it would serve as a point of reference and education for the whole of society, and it would establish appropriate mechanisms for enforcement. A charter of women's rights would accordingly give texture to the broad constitutional principles by focusing directly on questions of immediate and pressing importance to women, such as health, employment, reproductive capacity, violence against women, and child care.

One may therefore envisage a hierarchy of constitutional and legal provisions along the following lines:

- ☐ General principles of gender equality, non-discrimination, and affirmative action, to be found in the Bill of Rights section of the constitution. These broad formulations could only be altered by the relatively difficult processes of constitutional amendment.

- ☐ A charter of women's rights falling under the general umbrella of the Bill of Rights, but to be found in a separate document. The charter would have a comprehensive set of rights and remedies formulated in relatively specific form, covering the areas of employment, health, sexual harassment, and so on. In the light of experience gained in its implementation, and responding to the evolution of ideas and institutions, the charter could be amended more easily than the provisions of the Bill of Rights. At the same time it would have a special status as both a general code and as a point of reference for interpreting the Bill of Rights and for drafting relevant legislation. In other words, it would be entrenched, but less rigidly so than the Bill of Rights.

- ☐ Legislation adopted by Parliament or local authorities in keeping with the principles of the Bill of Rights and the norms and institutions of the charter of women's rights. These statutes would be eminently specific in character, and subject to amendment by simple majority according to the ordinary processes of Parliament.

- ☐ There could also be a series of enforcement mechanisms, such as direct appeals to the courts, the establishment of an equal opportunities commission to investigate complaints, court actions where necessary, or a gender rights council attached to Parliament with responsibility for reporting on the implications for women's rights of all proposed legislation.

Conclusion

The struggle to create a non-sexist South Africa will be even more difficult than the fight to create a non-racial one. It is not the constitution which will give women their rights, since, as has been pointed out, rights are never conferred, they are won. Yet the constitution could have an important role in consolidating the rights that women will have gained in struggle and in providing the basis for further advance.

Although the terms of the constitution will be important, even more significant will be the extent to which women are involved in the processes leading up to its adoption, and in the processes of its implementation. Yet the formulations do matter. They synthesize important ideas, serve as points of reference for those campaigning for improvements, and enter into the general culture of the nation. The constitution of the new democratic South Africa will be the product of a freedom struggle and should in its every clause breathe the spirit of liberty. All South Africans should see themselves reflected in it and protected by it. Even if the new South Africa is not described explicitly as being non-sexist, it cannot be called democratic if the voices of more than half the population are stifled.

The constitution should therefore be expressed in a language which makes it clear that it speaks for women as well as men; it should clearly affirm the equality in rights, dignity, and status of men and women; and it should create conditions which facilitate the right of women to express themselves and establish their own demands and priorities.

The general principles of the constitution should be enriched by a charter of women's rights focusing on all the concrete areas where the law and public policy play a role in affecting women's lives. The charter could in fact be drawn up before the constitution is drafted, and serve as the foundation for the clauses in the Bill of Rights dealing with the gender question. The precise relationship between the charter and the constitution is something that would have to be worked out, but clearly each would have to reinforce the other. The campaigning for and around such a charter would generate a consciousness which would go a long way to making the rights a reality and to reducing the pain and discomfort with which the subject is suffused.

6 The constitutional position of the family in a democratic South Africa

How can a society strengthen the family and at the same time weaken patriarchy? Nowhere in the world has this been fully achieved, yet this is precisely the daunting task facing us in South Africa. It is not impossible. We are living in a period of great social renewal in our country. Issues hidden for decades are now firmly on the agenda. The popular energy released by the struggle against apartheid opens possibilities of transformation in every area of public and personal life.

Apartheid has penetrated so violently and intrusively into the intimate lives of the majority of the people that only the complete elimination of apartheid laws and practices can permit anything approaching a normal family life to emerge. At the same time, the eradication of apartheid requires not simply the re-writing of obnoxious laws, but the repairing of millions of damaged families. The anti-apartheid struggle thus takes on a very concrete and deeply humane responsibility, that of helping to create the conditions for the pursuit of happiness in its most intimate and personal of forms.

We can and must theorize about the issues, but we can never forget that the great majority of us are involved in the matter under discussion on a daily basis. Nowhere is there such an interaction between the subjective and the objective, between the general and the particular, between what we say and what we do, as in the family. Nowhere are there more contradictions — courageous freedom-fighters who are tyrants at home, people who respond actively to the needs of the masses and yet deny that those with whom they share their most intimate activities even have needs (freedom-fighters during the day and fascists at night), and conversely, people capable of great tenderness in the family at night who are torturers by day.

And nowhere is the key to advance more evident — democracy. It is precisely because family life is so intimate and all-involving that the people themselves must be directly involved in the processes of its transformation. Happiness can never be imposed or decreed, not by the legislature nor by the church nor even by those whose lives have been committed to the pursuit of freedom. It has to be fought for and won by those who aspire to it. In determining the place of the family in a new South African constitution, the people must be involved at every stage — in determining priorities, in establishing general principles, and in administering the institutions set up for their implementation.

Our starting point must therefore not be some abstract, idealized model of the perfect family, but the actual lives that people lead today, within the general context of democratic transformation taking place in our country. On the basis of a hard look at South African socio-cultural reality we must provide the legal underpinnings for the resuscitation of family life in our country, but do so in a way that consolidates rather than undermines the general democratic principles for which we have been fighting. We need democracy in our processes, democracy in our mechanisms, and democracy inside the family itself.

The family has been grievously injured by apartheid; overcoming apartheid means, amongst other things, retrieving the family from the depths of its trauma. At the same time, apartheid has been particularly devastating to the rights of women. Dismantling apartheid therefore requires that special attention be paid to undoing the many laws and practices that seek to keep women subordinated. To restore the family in such a way as to constitutionalize male tyranny, whether benevolent or brutal, would be to eliminate one of the effects of apartheid while strengthening another. It would be deeply undemocratic: in the first place, it would deny more than half the population the right to decide what kind of family law they should have, and secondly it would suppress the wishes of half the family partnership on a day-to-day basis.

From slavery to apartheid

The first point that needs to be made about the damage done to the family in South Africa is that it occurred not simply as a marginal or indirect consequence of the process of industrialization and urbanization — as happened in so many other countries — but as a result of deliberate policy and calculation.

In the Dutch slave-owning settlement at the Cape, there was explicit negation of the family rights of slaves. Slaves were possessions of others, not possessors of rights themselves. Their indigenous family law was completely shattered, and they were in general not admitted to baptism and the church, and so excluded from the family law of their masters and mistresses. Thus their unions were not recognized as legal marriages, they could not own their own homes, or be heirs to property. They did not even have their own names.

In the later colonial period, the attack on the family took a different form. In traditional African society, the household had been the foundation of political and economic life. In order to uproot the people, to destroy their independence, and detach them from the land, the colonial authorities felt it necessary to attack the household and disrupt its self-sufficiency. The weapons were ideological (African customs were called savage, African beliefs heathen), legal (head and so-called hut taxes were imposed), and economic (the monetarization of all relationships, including lobola or bride-wealth).

The tax system, the pass laws, the establishment of compounds for mine and farm labour, the creation of what were called black locations on the periphery of the urban areas, were all designed to split African families and compel the menfolk to work for whites on the basis of single-person wages, while the womenfolk produced new generations of labourers in the so-called reserves. The mines and large parts of farming and industry became dependent on migrant labour; the political system of apartheid was little more than the superstructure of migrant labour. Family life for Africans was to be made impossible in the reserves and illegal in the towns.

The splitting of families thus became deliberate policy enforced by law. The Native Labour Regulation Act provided for the establishment of single-sex compounds in which no family life was permitted; the much-hated Section Ten of the Native Urban Areas Act was specifically designed to prevent African women and children from living with their husbands and fathers in the towns. Some of the more lenient magistrates in the rural areas actually allowed African women to spend a few weeks with their husbands in the towns for what the authorizing document called 'biological reasons'.

The wider legislative context for the transformation of family law

The restoration of African family life thus has relatively little to do with family law and very much to do with the general structures of apartheid law. Any serious attempt to promote respect for family life in South Africa must deal concretely with the laws and practices associated with the migrant labour system. Employment practices have to be revised, wages paid on a different basis, compounds progressively and rapidly phased out and replaced by family homes, and the rural areas rehabilitated so as to become self-sufficient once more. Greater access to land becomes vital.

The crucial element will be the involvement of the people. The migrant labour system cannot simply be banned, it must be transformed in an active and coherent way, with trade unions and neighbourhood organizations playing a key role. The interests of neighbouring countries would have to be taken into account.

One direct outcome of the migrant labour system was the denial for decades that Africans living in the towns had a right to decent housing. Thus today we have homelessness on an enormous scale. The lack of housing is doubly injurious — it prevents any possibility of stable and decent family life, and it is a massive reminder of social inequality. The immense inequalities created by apartheid have to be removed by active programmes of affirmative action. As a first step, every home should be guaranteed accessible safe water, then furnished with clean piped water and electricity or gas. The true hewers of wood and drawers of water in South Africa are the millions of women who lose hours every day on survival tasks that should be shouldered by society as a whole.

Thirdly, the whole of social security, tax and income maintenance law must be re-formulated. These have been based on the claim that the African family subsisted in the reserves outside the norms and expectations of modern industrial society. Such social security as existed was miserly for blacks compared with whites, and to this day, social security law leans heavily towards benefiting most those who already have most. Thus, pensions for whites are considerably higher than those for blacks; whites receive far more protection from unemployment insurance and workers' compensation law than do blacks. The African people have accordingly been denied their traditional family-based support systems and at the same time excluded from the mainstream of state-based social benefits.

When people finally take control of their own lives and are free to determine how and where and with whom they shall live, many questions which today become tangled with the defence of apartheid can be faced on their merits. One of these which has special importance for the quality of family life is the control of fertility.

Today any government-sponsored programme aimed at family planning is seen by many as a device to keep the black population as low as possible. The fact is that there are a huge number of unwanted pregnancies that could have been avoided with access to birth control. Similarly, the hospitals are filled with patients haemorrhaging after clandestine abortions. These are sensitive questions in which cultural beliefs and religious convictions play an important role for many citizens. What is required is honest dialogue on the subject. The issues have to be brought out into the open and discussed calmly and objectively.

When we speak of the people, we are not speaking of a monolithic mass, but of a large population with the most varied experiences of life and the most diverse range of views. Reactionaries will have the right to voice their opinions along with everybody else — indeed, it is better that they subject their ideas to debate than that they resort to sabotage and deception to get their views across. Frequently on social issues of this kind, the only consensus that can be arrived at is an agreement to disagree. What becomes vital then is that the law and social practice tolerate a variety of opinions.

Those who are against birth control or against abortion will have the right to argue their views and work towards finding alternative approaches, but will not have the right to impose their positions on others who hold different opinions. Similarly, those who favour contraception and the right to terminate unwanted or dangerous pregnancies should be free to put forward their positions but not have the right to insist on birth control and abortion for those who do not want it. What apartheid society has never done is allow people to choose for themselves how they wish to lead their lives. What post-apartheid society must do is guarantee to people for the first time the basic rights of personal self-determination.

In many cases, these are questions which will be discussed within the family and agreed upon by those affected. There will unfortunately be cases where, say, the partners to a marriage cannot agree. For example, the wife may wish to use contraception or to terminate a pregnancy and husband may oppose her, or children might wish to use contraception against the orders of their parents, who may see contraception as an invitation to immorality. There is not much that the law can do about these situations, save in the ultimate instance to recognize that no one should be forced either to conceive or to carry a foetus against her will (certainly not in the early months of pregnancy), nor, by the same token, should anyone be obliged to accept involuntary contraception or abortion.

The state should be obliged to provide facilities to permit the free and informed exercise of choice. On the one hand there should be a range of counselling, support, and adoption services (where wanted) for those who wish to carry their pregnancy to term. On the other, there should be access to hygienic and dignified abortions for those who wish to terminate their pregnancies. The criminalization of abortion should be ended, since apart from moral objections, its only consequence is to crowd the hospitals with patients suffering from the effects of badly conducted illegal abortions.

Easy access to contraception is another right people should have. Apart from enabling spacing of children, contraceptives can play an important role in stopping the spread of sexually transmitted diseases, more especially AIDS. At the same time, the neglected question of infertility, responsible for much sadness, requires special attention.

Lastly under this heading, there is a great need for sex education, not only amongst the young. Sex education in the full sense connotes much more than biology lessons on how the birds and the bees do it. It deals with human relationships at their most intimate, and raises questions of responsibility, tenderness, trust, and respect, as well as fundamental issues of how males and females relate to each other, and for that matter, to members of their own sex.

Another area of apartheid-induced neglect is that of health care, particularly in relation to mother and child. Pre-natal, maternity and post-natal care for the mother, and immunization and basic nutrition for the child are the very foundations of a secure family life, yet their provision for black people remains extremely scanty both in the towns and the countryside. Massive infant mortality in a country that pioneered heart transplants is totally unacceptable. There can be no repair of family life without eliminating the pain and tragedy of unnecessary infant death.

Finally, virtually no provision has been made for crèches, kindergartens, and schooling for black children, who again find themselves denied both the sustaining force of the traditional extended family and the support of institutions normally associated with industrialized societies. Similarly, school-feeding for African children was stopped when white farmers complained that this was spoiling the children and making them lazy. Unequal access to education continues all the way through, putting black families under enormous pressure to achieve educational opportunities for their children which white families simply take for granted.

Thus restoring family life in South Africa requires far more than simply looking at and renovating family law. At the constitutional level it necessitates clear principles, and at the legislative level clear programmes aimed at removing all the myriad ways in which apartheid damaged family life and created intranquility, poverty, and a sense of dispossession for the mass of the people.

Transforming family law

There is no such thing as a typical South African family, let alone an ideal one. There are many South African families, constituted and dissolved according to a great variety of marriage and divorce systems. The varied origin of the people who make up the population of our country is reflected in the multiplicity of marriage rites. We have marriages based on lobola or *bohadi*, marriages solemnized in church or temple or synagogue or before the imam, and marriages celebrated in civil registries.

The same couple could marry three times — with lobola in the traditional way, then at church and thirdly, if the church is not recognized by the state for purposes of registering marriages, before a civil marriage officer. There are also many people living in stable unions, constituting families, without any process of formal solemnization. Finally, a very large part of the population live in single-parent families, occasionally through choice, but usually through abandonment, widowhood, or divorce.

At present, the law does not give equal recognition to all the different kinds of marriage. Marriages celebrated before a state marriage officer or before a religious officer recognized by the state as a marriage officer receive the fullest recognition. They are registered and marriage certificates are issued which identify the parties, state the date of the marriage, and have strong evidential value in a court of law. The legal consequences of such marriages are in general laid down by statute or by the common law as interpreted by judges, with the parties having some say over the kind of property relationships that are established by the marriage. The law stipulates certain prerequisites for such a marriage to be valid, the most important being free consent, minimum age, and the non-existence of another marriage to which either the bride or the groom is a party.

Traditional marriages are also given a measure of recognition by the courts. The rules governing these marriages are derived from what is referred to as customary law, that is, unwritten law passed down from generation to generation. In fact the status of this law today is highly confused, since the law as applied in the courts has been heavily influenced by decisions over the decades by white magistrates and judges as well as by white text-book writers. The law is thus written and unwritten at the same time; it belongs to the people and yet no longer belongs to them. Grave evidential problems arise as to proof of the existence of a marriage — was sufficient lobola paid, what exactly was the understanding between the families, what was the moment when nuptial negotiations turned into a legally binding union?

Other criticisms have been offered: far from lobola today serving to bring two families together, in contemporary conditions it commercializes marriage and gives rise to endless disputes; the customs are ethnically based and encourage a sense of ethnic apartness rather than national identification; in certain rural areas, corrupt and unpopular chiefs manipulate family law for personal advantage or else to help their favourites; the rules themselves are frequently out of keeping with the way African family life has evolved, for example in relation to minimum age, polygamy, voluntariness, the treatment of adult women as though they were minors, the awarding of children to the father's family if lobola has been paid, even if they would be better off with the mother.

To complicate matters, during the Verwoerd era, apartheid officials concocted something called Bantu law as an ideological cover for excluding Africans from a common society. The most archaic and authoritarian aspects of traditional law were emphasized, while the democratic features were suppressed. In traditional society, as in pre-capitalist society on other continents, family law was public law, the governing class succeeding to authority according to the rules of family lineage. The Verwoerdian scheme was to offer a spurious, tribalistic family law as an alternative to democracy and the vote.

Yet, without doubt, the traditional customs regarding marriage still have great significance for people, connoting respectability, seriousness and sensitivity to the culture of the community. It is 'our way' of doing things. People regret the debasement of tradition rather than challenge its existence.

Special rules exist with regard to the recognition of Hindu and Muslim marriages in South Africa. For decades their status was a source of fierce conflict. The official position was that since these unions were potentially polygamous (even if actually monogamous) they could not be regarded as true marriages. Wives were accordingly treated by the law as concubines, while children were regarded as illegitimate. Eventually, in the face of considerable mobilization by the communities directly affected, the authorities acknowledged by means of special legislation that Hindu and Muslim marriages were indeed marriages entitled to recognition in law.

The same cannot be said about marriages celebrated in independent African churches, nor about unions of a long-lasting and stable kind constituting families recognized as such by the community, but without having been preceded by the formalities and rites of either a registered marriage or a traditional one. In the absence of properly researched statistics, one can only guess at the number of such families, yet it may well be that they are as numerous as the combined total of all the registered marriages and marriages in which the full procedures of traditional law have been followed. The law has tended to frown upon these common law marriages or *de facto* unions, as they have been called, generally treating them simply as cases of co-habitation outside family law and virtually outside any legal framework whatsoever.

The problem in a democratic South Africa will be how the law and the constitution should regard this great variety of marriage systems. Registered marriages are non-racial but not particularly democratic. Traditional marriages are popular but certainly not non-racial. Millions of people live in families that the law does not recognize as such. Should there be a single legal regime of marriage, the same for everybody irrespective of background, culture, or preference, or should there be a legally recognized plurality of marriage systems? There are a great number of options, and many nuances within them.

One radical possibility is to have a unitary system of marriage, recognizing only one form of marriage rite and denouncing all others. This has happened in theocratic countries where religious fundamentalism has monopolized marriage law, and also at certain moments in some post-revolutionary societies. Theoretically in South Africa there could be a single form of marriage recognized by the state, one which say, emphasized non-racialism, national unity, and equality between the spouses, coupled with state action to denounce religious marriages as superstitious and traditional marriage customs as tribalistic and feudal.

A softer and less intolerant variant of this policy would be to have a single South African marriage law based upon a single concept of the essential characteristics of marriage, but enabling various rites to constitute due solemnization. Various state, religious and community leaders could be recognized as marriage officers capable of performing or recognizing marriage ceremonies. They would have to satisfy themselves that the pre-requisites of a proper marriage were present (for example minimum age, monogamous relationship, free consent) and make some form of registration, but otherwise the ceremonies could be in a magistrate's court or a church or a temple or a synagogue or in a village centre or at a homestead, and accompanied by prayers or the slaughter of an ox, and in Zulu or Afrikaans or Tsonga.

What would matter would be that, irrespective of the form of marriage followed, the law would attribute the same rights and responsibilities to the couple, with possibly some choice regarding property relationships. Beyond this, the parties would be quite free, if they both so wished, to apply the particular rules of their faith or custom to the marriage. Thus persons married in a Catholic church might accept that their marriage is indissoluble, even though the law granted them the possibility of divorce. Similarly, if lobola were paid, the intricate rules governing the relationships between the two families involved might be followed in detail, if the persons concerned so wished.

These social and religious rules would be enforceable according to the convictions of the parties, and to some extent according to community pressure, but not in law. The marriage law would establish a common set of fundamental principles applicable to all recognized marriages, principles which could be invoked by either party, with or without the consent of the other. To sum up: the religious or traditional rules would operate outside the formal legal system and have sanctions of a moral and social but not of a legal kind.

Such an approach has many advantages. It encourages the concept of a common society, with a common citizenship and a common platform of legal rules applicable to all, irrespective of colour, language, religion, origin, or gender. Family law would be set in the context of fundamental constitutional rights that emphasized the basic principles of democracy, freedom, and equality.

At the same time it would be flexible and sensitive to the cultural and religious diversity of the country inasmuch as it would acknowledge that there are many ways in which people like to marry, and it would be tolerant in the sense that it would permit informal marriage rules based on tradition or religion to exist outside the formal state sector. Individuals and families could continue to follow practices and beliefs that have special meaning for them, and the law would only intervene if they could not agree amongst themselves and at least one of the parties preferred to invoke his or her constitutional rights.

Yet there are problems that have to be faced. There are, for example, certain communities that may refuse to have their marriages registered if this automatically imposes a set of constitutional rights and duties on the parties. Such communities may further object that if the state courts are not allowed to resolve disputes according to the rules of tradition or the Holy writings, then people will simply boycott the official court system and have recourse to mechanisms of their own. This latter point can be met to some extent by transforming the composition of the judiciary, so that it becomes more representative of the population as a whole, and by attuning court proceedings more to the local cultural setting. This theme will be developed separately.

The only way to resolve these questions is to discuss them directly with those whom they affect most intimately. The issues are not merely symbolic or cultural. They affect pension rights, rights of succession, questions of custody, and of division of property. They also touch on the status and sometimes the income of traditional leaders. In some cases they deal with the very concept a community has of itself.

Thus the Muslim community has a specific cultural identity bound up with subjection to slavery and later, to their status as indentured labourers. The struggle for the maintenance of Muslim tradition, including Muslim family law, was part of a struggle against racist domination, which frequently took the form of aggressive and hegemonic Christianity. The Muslim community must articulate its own views on how best to combine the twin goals of creating a non-racial, non-sectarian democracy based on the principle of equal rights and duties, and preserving and developing the cultural and spiritual heritage of all the various communities, including their own, that make up the evolving South African nation.

The same point can be made about sectors of society in which the defence of the institutions of traditional family law was part of defending the community's integrity in the face of colonial invasion and apartheid dispossession. There may be areas of the country where the people actively involve themselves in the general democratic struggle and accept the broad principles of national unity, yet at the same time wish to preserve traditional rules and practices in relation to family law. This can only be discovered by means of open debate. The point is not so much to put general philosophical questions to people, such as: 'Do you favour the continuation or abolition of traditional law?', but rather to discuss the concrete options available, and in particular to see what degrees of compromise or transitional arrangements are possible.

Thus in principle it should not be impossible to have a single system of basic rights and duties attaching to all marriages, whether civil or church or traditional, which would be recognized and applied by the courts, and at the same time permit the establishment of conciliation machinery outside or alongside the court system, which at the request of the parties, could give more weight to traditional or religious norms.

The above are all variants of what has been referred to as a unitary system of family law. The new constitution could, however, reject any attempt to create a single marriage law for South Africa, and opt for one of the many variants of legal pluralism available.

The most radical pluralist solution would be to regard family law as being determined by the personal law of each couple, in terms of which there are neither common norms nor common forms of administration. Thus persons married according to traditional law would have their cases judged by traditional judges applying traditional rules; each religious group would have its own judicial figures who would decide on family disputes involving members of their congregations — canonical courts, rabbinical authorities, Muslim judicial councils; and the state courts would only have jurisdiction in the case of persons married according to state marriage legislation.

A less radical version would be to have a single state system of justice responsible for the administration of family law, but to permit the judges to apply the principles and rules of the marriage system most relevant to the case.

This is what was done in Tanzania, where the further step was taken of establishing certain minimum ground rules to be applied in all cases, irrespective of whether the marriages were civil, religious, or traditional. Thus a national minimum age was established, and also the principle that a divorce only became legally effective when made part of a court order.

Accordingly, the courts would recognize Muslim marriages, even if actually polygamous, and also divorce in terms of the Koran by *talaq* or repudiation, but the divorce would only operate legally from the moment the judge recognized the existence of the *talaq*. If the marriage was accomplished by means of lobola, the court would apply the rules of lobola to the situation, save that minimum age would have to be respected, and the divorce would have to be decreed by the court and not by any traditional leader.

In the case of a pluralist system administered by a single judiciary, one of the key questions becomes the determination of the legal norms applicable, the so-called choice of law problem. Thus the first question the court would have to decide would be whether to regard a marriage as civil, religious, or traditional and which legal doctrine would apply in each case.

Put more concretely, the basic choice is between a single system of rules and administration (that tolerates an informal multiplicity of marriage procedures outside the formal system), or a variety of separate systems within the overall legal structure, each with its own rules, and possibly also its own administration.

In practical terms, it means that judges would respond to a particular case in one of two ways.

They could say: this is a Muslim couple, therefore we apply Muslim law; or this is a Xhosa marriage, therefore we apply Xhosa law; or these people were married in the Catholic Church; or are Jews or Jehovah's Witnesses or African Zionists and we must apply the appropriate rules of each religion; or they were married before a magistrate therefore we apply state family law provisions.

Alternatively, judges could say: whether they are Muslim or Xhosa-speaking or Catholic or Jews or Jehovah's Witnesses or African Zionists is their business; they have freedom of religion and the right to organize their family life as they wish, subject only to restrictions against domestic violence, child abuse, and so on; they can try to resolve their problems by resorting to traditional or religious leaders, or just accepting the decisions of the family councils, if they so choose; but once they bring the matter before us, we will apply the general principles of the new South African law, irrespective of their religious or ethnic background.

If the latter approach is accepted, it is possible to lay down certain general principles which the courts would apply, although these too would have to be thoroughly discussed by the people before being incorporated into the law. There has over the past three decades been a tendency at the international level towards the universalization of certain family law concepts (though the revival of the *Sharia* as the source of family and criminal law in a number of Islamic countries represents a counter trend).

There has been a general move in most parts of the world towards prohibiting child marriages, encouraging monogamy, insisting on voluntariness as the foundation of marriage, defending the principle of shared parental responsibilities and rights in relation to children, accepting equal rights and duties between the spouses, and acknowledging that on the breakdown of the marriage the family home should be disturbed as little as possible and that property acquired during the marriage should be shared equitably, independently of who actually paid for it. All these principles have already taken strong root in South Africa. The problem would not so much be how to state them as how to apply them.

Mechanisms for implementing the new family law

If a policy of full legal pluralism is adopted, there will be a multitude of judicial officers applying a multitude of different rules. The question of how different kinds of judges are selected could become quite complicated, each sector having its own set of qualifications and mode of deciding who should exercise judicial power.

On the assumption, however, that some form of unitary administration of family law is going to be adopted, one may advance some ideas as to how judicial structures can be transformed to make them more democratic and culturally sensitive.

In the first place, the whole of the judiciary will undergo major changes to make the bench more representative of the people as a whole. The judges in post-apartheid society will have an important role to play in defending the constitutional rights of citizens. For people to have confidence in the judiciary, it will be essential that they see themselves and their highest qualities reflected on the bench. The idea of one section of the population sitting in judgement on another will have to go. This means that the judiciary will be composed of persons of integrity and skill representing the wisdom and humanity to be found in every section of the community. Not only will the judiciary cease to be a white preserve, it will lose its male-only character. With the development of a new language policy, African litigants should be assured of the right to have proceedings conducted in the language with which they feel most comfortable. The involvement of lay assessors and possibly juries will ensure that the gap between litigants and adjudicators is diminished.

These are general transformations that would favour the resolution of family disputes in a more just and sensitive way. In addition, specific attention would have to be given to the creation of a system of Family Courts, common to many countries, but with a special South African flavour. These courts would be part of the general court system, but operate in a manner appropriate to their particular competence. Thus, within a framework of common constitutional and legislative norms, they could have a considerable degree of flexibility in the way they function.

In certain rural areas, they could be composed, in traditional manner, of several members. Their procedures could be informal and largely oral in character. In this way the vitality and flexibility of traditional methods of resolving family disputes could be maintained, while introducing new elements, such as having women on the bench and applying principles based on equal rights between the parties.

In urban areas there could be more emphasis on formal proceedings and legal representation, yet even here there would be scope for infusing the family courts with the democratic aspects of African tradition. Community courts with jurisdiction to deal with family problems, neighbours' disputes, and relatively minor breaches of the peace, could ensure an appropriate combination of legal rigour and social informality. The rules would be the same as for any other court, but the atmosphere would be one in which any member of the local community would feel at home.

For those sections of the population not served by the abovementioned courts, there would be an additional network of family courts comprising persons of all backgrounds. These courts would not be too different from the civil courts operating today, save that they would be more representative and more flexible, in keeping with a world-wide trend to look at family matters 'in the round' and less as subjects for litigation in the formal legal sense. Thus certain basic constitutional procedural rights would be guaranteed, while there would be less emphasis than at present on strict rules of pleadings (documentary statements of the issues) and on technicalities of how evidence can be adduced. The provision of legal aid can be of special importance in family law matters, especially to the parties, mainly women, whose financial position is weaker.

Finally, there should be ongoing monitoring of the way family law matters are handled to ensure that the law and the procedures evolve satisfactorily. As far as particular cases are concerned, the only remedy for an aggrieved party would be to appeal to a higher court. Yet trends could be noted, and tendencies systematically to flout the new constitutional principles could be brought to light and counteracted.

In the end, the efficacy of the law will depend upon many extra-legal factors: the degree of general public consciousness, the vigilance of women's organizations, the scrutiny of the press, and the way in which the judiciary succeeds in implanting itself in the community while maintaining its independence. Yet the terms of the law and the way the courts function will be of major significance. It is essential that attention be paid to these questions now. The struggle for a new family law is part and parcel of the struggle for new family relationships and for a new South African nation.

7 Towards a charter of children's rights

In both classical law and traditional African law, the basic rights of the child were essentially restricted to the right to a name and the right to inheritance. In recent years, the right to support, always strong in African society, and the right not to be abused, have been added. Clearly, in a democratic South Africa, all these rights will be preserved and strengthened; indeed, the removal of apartheid is a pre-condition for their large-scale realization. Yet to restrict the right of the child to these narrow areas would be to turn a blind eye to the true deprivations imposed on the child in apartheid South Africa and to ignore the full range of the children's claims.

A narrow concept of children's rights is appropriate in relation to one particular set of rights, namely those enforceable through the courts against cruel or neglectful parents. Since it is not the state that creates the family and family relationships, great sensitivity must be applied in this area. But the situation is quite different when the neglect and cruelty come from the state, and when parents themselves are oppressed, whether directly by the domination of the apartheid system or indirectly as dominators themselves, their minds imprisoned by terror and hatred. Then children's claims go beyond being merely claims against their parents enforced by child protection societies or officers of the state. Their claims become claims against the state itself, requiring appropriate legal guarantees and enforcement mechanisms.

The greatest abuse to which South African children are subject today comes from the organized might of the state. Any charter of children's rights in a democratic South Africa has to take this fact as a starting point. The question of children's rights thus cannot be separated from the general struggle to eliminate apartheid. At the same time, it is a question that has its own particularities. The struggle for children's rights contributes towards and enriches the general struggle against apartheid. It is also a struggle that will continue after apartheid is destroyed; in fact, it will only fully achieve its objective once the country has been liberated.

The first step must be to restore to the children everything of which the apartheid system has robbed them. This means recognizing claims, both of a moral and a legal kind, that can be enforced either directly by children or else on their behalf, against society as a whole and all its institutions.

The right to grow

Every child has not merely the right to live, in the sense of the right to survive, but the right to grow, to develop his or her physical and mental capacities. Apartheid society denies this right to the great majority of South African children. Hunger in a country of wealth, and social diseases such as TB and kwashiorkor in a land of advanced medicine are proof of this. Cold actuarial statistics show that black infants have a first year mortality risk twenty times greater than that of whites; that on average, every black citizen has a life expectancy twenty or more years below that of whites. The black children of our country, like the white, have a claim on the state and on public and private institutions to create conditions for generalized primary health care, including mother and child protection and for guaranteed minimum nutrition. While non-governmental organizations have a valuable and continuing role to play in these areas, the issue cannot be left on the moral or altruistic plane alone or to the goodwill and spare time of volunteers. The right to altruism is an important right which should not be submerged in the new South Africa. But it should never be seen as the main guarantor of children's rights. Children have not only moral claims on society, they have legal ones as well. Legislative programmes are required to establish clear goals, set out the minimum food and health-care requirements of each child, and provide a statutory basis for the progressive achievement of these goals. These programmes must also provide a statutory basis for their progressive achievement. Once apartheid has been removed there will be no impediment to the creation of a legally-based system of child welfare that materializes in tangible form the right of the child to grow.

The right to play

Argument rages in many countries about whether the sale of war games should be lawful or not. In South Africa, the problem is not whether children should be encouraged to fantasize by playing with model tanks or rifles, but whether real armoured cars, automatic rifles, and teargas-throwers should continue to dominate their lives. When children dodge in the streets, it is not to escape imagined 'cowboys or crooks', but to evade real killers frequently bent on murder. When their school grounds are occupied by troops, when their courage is displayed not on the sports field but in the torture chambers of the police, when persons acting in the name of law and order are licensed by indemnity to kill at will, when children are slaughtered in their houses while vigilante gangs act with the connivance of the authorities, then it is clear that the law has been converted into an instrument of lawlessness and that the games children play become literal games of life and death. When children report on fatalities — 'our side' versus 'their side' — as if giving a football score, we all bear the shock. In a democratic South Africa, the law should defend especially the children, and impose duties on national and local authorities to provide properly supervised facilities for them to enjoy sport and recreation. The use of violence against children whether in the home or the school or in the streets, should be prohibited by law, and those responsible for such violence severely punished.

The right to learn

The South African statute book is filled with legislation dealing with questions of education, but the main objective is not to ensure that education is guaranteed but to guarantee that education is separate. As the late Sir Robert Birley pointed out, in every other country in the world, education is used to promote, at least at the surface level, a sense of common nationhood and equality; only in South Africa is it used to promote disunity and inequality. The first educational right children should have, therefore, is the right to learn together. As the US Supreme Court declared in the famous case of *Brown v Mississippi Board of Education*, segregation is in itself discriminatory and harmful. As subsequent USA experience has shown, formal desegregation in not enough. An active programme of affirmative action, binding on the state, on public authorities and on the schools, is required to convert abstract legal rights into social reality. State schools, private schools, and church schools all have a role to play in encouraging, through their composition, their practice, and the content of their education, the notion of an undivided South Africa inhabited by free and equal citizens. Children have the right to study in their mother tongue but also the right to study in other languages if this gives them greater access to world culture. Children have the right to truth, about themselves, their bodies, who they are, where they come from, about the world they live in, and they also have the right to know that the truth is often complex and always filled with contradiction. One cannot legislate truth, but one can legislate for conditions which promote the truth, and one can ensure that information comes not from a bureau but from experience and life itself.

The right to adventure

One of the greatest and most elusive of all the rights of the child is the right to adventure, the right to explore one's environment and in so doing explore the limits of one's body and mind. In South Africa, the law bars this right to the great majority of children, so that the only significant adventure permitted to them is to challenge the law itself, with all the terrible consequences that follow. In a democratic South Africa, conditions will be created for lawful adventure. The country will belong to all who live in it — the mountains, the rivers, the beaches and parks — will be open for all to explore. Programmes will be established with a statutory basis so that all children, and not just a minority, learn to swim; so that all have access to the pleasures of cycling, mountain-climbing, and camping. (Today white kids live in a tent for fun, black kids because their home has been bulldozed.)

The right to imagination and culture

Children in apartheid South Africa grow up not only in physical ghettos but with ghettos in their minds, cut off from each other and severed from the ideas and culture of the world. They are permitted to know next to nothing about their own continent, its history and culture, and what little they are told is distorted and pernicious. The majority of children have been informed by all kinds of direct and indirect means that they are diabolic, that their traditional culture is debased, and that their demand to wander in what has been called the green pastures of the enlightened minority is sinful, unnatural, and unlawful. Instead of the great diversity of cultural sources in our country being a foundation for richness, vitality, and interchange, it has been converted into the basis of enmity, suspicion, and domination.

The law, for so long an instrument of brutalization that suffocates the imagination, needs to be completely transformed. It must not only lose its present negative character ('If you don't behave, I'll call the law'), it must become a positive juridical bulwark of creativity, guaranteeing freedom of information and access to ideas and outlawing the preaching and practice of apartheid and disrespect for others.

The right to warmth

Every child has the right to be cherished, to grow up in an atmosphere of warmth and security. In present-day South Africa, the law, acting through bulldozers and armoured cars, crushes the emotional security of the majority of children, while migrant labour, tied in with the bantustan system and the apartheid control of housing, tears families apart, denying children stable homes and constancy of parenthood. Children see their parents humiliated and insulted by soldiers and police, and are themselves direct victims of state arrogance. Nowhere are they safe, not in their homes, not in their beds, not in their schools, not in the streets, not even in church or mosque or at the graveyard.

The law as such can never guarantee love and security, but it can be a major instrument in promoting conditions which favour the achievement of these goals. It can provide for a network of family support agencies, impose duties on local authorities and employers to pay due respect to family situations in relation to housing and employment, and require the establishment of crèches and kindergartens for children of working parents.

The right to worth

The right to worth and dignity should be one of the most respected rights in law, but is one of those most denied in apartheid South Africa. For many purposes, many South African children do not even rise to the level of statistics — births are frequently not registered, nor in many cases, are infant deaths. There are no precise statistics for black infant mortality, nor for the life expectancy of blacks, nor for the rates of malnutrition; this, in a country where the whole adult population is fingerprinted, photographed and on file, where highly sophisticated systems of control have been established. At best, children have been seen not as the bearers of rights, each with his or her own personality, each destined to be a full citizen in the land of his or her birth, but as future labourers, bureaucrats, or policemen, their lives perpetually at the command of the apartheid rulers. That millions of children should go barefoot and in rags in a land of plenty is not regarded as scandalous. The total absence of legislation guaranteeing minimum rights for children flows from concepts in which the majority of children just do not count.

In a democratic South Africa, starting with the constitution, the law will recognize the special duty of society to cherish its children and enable them to lead lives in which they experience their worth and develop the confidence and conviction they need to be free and active citizens in a free and developing country. The state will be obliged to acknowledge the identity of every child, recognizing the equality of all and the dignity of each. Distinctions based on colour, origin, or parentage will be outlawed and all documents issued by the state shall serve as guarantees of rights and not as instruments of oppression.

The rights to fellowship

One of the major crimes committed by apartheid against the children of South Africa has been the fostering in their ranks of division, enmity, fear, and contempt. Instead of encouraging a sense of sister- and brotherhood, apartheid put child against child. White children have been taught to deny their humanity, not to see their black compatriots except as servants to be exploited or enemies to be controlled. Black children have been treated by the system as if they had no humanity at all, as if their only future was to be humble servants to the oppressors or else compliant, corrupt, and cruel collaborators against their own communities. Children have been divided not only along racial lines, but on linguistic, tribal and, more recently, people-versus-collaborator lines. One of the major objectives of the state of emergency was to try to crush children's organizations and children's leadership. The detention and torture of thousands of very young people were testimony to this.

In a democratic South Africa, not only will the whole constitutional order encourage maximum human solidarity between children, but there will be a statutory basis for self-organization by children, so that they will have the right to associate freely and express their wishes and demands, whether through school councils or community children's groups or the transformation and development of existing children's organizations (such as the scout movement) into genuinely non-racial and democratic organizations true to the highest principles they espouse. The multi-cultural character of South Africa will be respected not by segregating children into separate organizations, as at present, but by encouraging the maximum interchange coupled with the greatest possible cultural sensitivity. The earlier the habits of mutual respect and solidarity are acquired, the sooner the nightmare of apartheid will be over.

The right to enjoyment

South Africa is a country of immense resources and has zones of great natural beauty. Life could and should be enjoyable for everyone. Yet, especially for children, life is an inferno. The insult and threat that apartheid represents, the hunger, the lack of decent clothes for the majority, the absence of secure homes, coupled with apartheid's physical brutality, stifle natural enjoyment of life. The law becomes an instrument of terror instead of a guarantee of tranquility. Children have to be perpetually on the run, they are turned into wanderers and scavengers, they have no sense that the country, the whole country, belongs to them. Their pleasures are few, and more often than not, illicit.

The elimination of apartheid laws and the complete dismantling of the apartheid system are the basic pre-conditions for children to exercise their right to enjoyment. The ending of repression and violence against children will follow as a natural consequence. But more will have to be done than simply to remove these evils. The law will have to play a positive role in overcoming the effects of past discrimination and enabling life to be equally rewarding (and felt to be such) for all children. Whether it is called positive discrimination, or affirmative action, or corrective anti-apartheid measures, the law will have to express itself in an active and dynamic way if this right is to become a reality.

The right to childhood

Inspiring though it may be to see how many children of our country have shouldered the responsibilities of adults, it is also horrifying that in the process the present generation should pay the price of being robbed of its very childhood. The slow accumulation of experience, the ability to fantasize freely and to have fun with the body and mind, all the spontaneity of infancy and early youth, are denied when children have to confront the barbarous physical presence of the apartheid state, to dodge teargas and bullets, resist torture, and engage in sanguinary conflicts.

A free and democratic South Africa will, for the first time since colonial domination and apartheid were imposed, guarantee the right of children to be children, to achieve adulthood in their own good time.

The right to a future

Hailed as heroic young lions by many, and denounced as precocious tyrants by others, the children of our country have undoubtedly borne the brunt of the anti-apartheid conflict. Excessively street-wise and prison-wise and combat-wise, they now have the right to emerge from their trauma and claim the future for which they have struggled so hard.

Once the slogan was: liberation now, education later. Today it has to be education now, liberation now. Tens and hundreds of thousands of specialists will have to emerge from the ranks of the oppressed if the country is ever to know true equality.

The present generation of young people have the right not to be fossilized as premature veterans, nor to be converted into eternal victims of a vicious society, nor to be everlasting exhibits to the cruelty of apartheid. They have the most fundamental of rights, the right to have rights, and also the most fundamental of responsibilities, the responsibility to be responsible.

We of their parents' generation have our rights and responsibilities too. We cannot use the notion of living for the future as a pretext for evading our responsibility for the present. We fight for a free and democratic South Africa because that is what we want for ourselves. It is less selfish to acknowledge that fact and live our lives in an open and spirited way, than to appropriate the future in a pre-designed scheme and then complain later of our children's ingratitude for not accepting what we did for them. Our sacrifices have been for ourselves, not for future generations. Our major contribution is to create conditions to enable our children to have their own future. Their right to have rights implies that they are not victims of history; their responsibility to have responsibilities requires them to assume their potential as full human beings now, sensitive to who they are and caring about what they do.

The children of our country should accordingly not be our adjuncts to be mobilized for this or that action and then told to shut up and wait for the future we are organizing for them. Nor should they be regarded as free-floating bands roaming the streets in constant tension with the community. They have the right to be free citizens of a just society and not angry victims of an unjust one.

The greatest of legal rights they can have is to grow up under a new constitution that recognizes the worth, dignity, capacity, and yearnings of all. If we are the mothers and fathers of that new constitution, they are its daughters and sons, and we are all its authors.

Towards an appropriate strategy of guaranteeing children's rights

For lawyers to whom the right to sue your neighbour is the basis of all legal rights, the idea of a charter of children's rights might seem more like poetry than law. The problem really lies with such lawyers, and not with the charter; they must open their eyes to the new range of legal strategies developed in recent decades in various parts of the world. In particular, the adoption by the United Nations of the International Convention on the Rights of the Child, provides a secure foundation for legislation in all countries.

Apartheid has thrived on a tight and technicist concept of the law which demotes or even completely excludes the human and social dimension. Such an approach was well suited to the use of law as an instrument for bringing order to the affairs of the minority and keeping the majority in their place. It is quite inappropriate when law is regarded as a means of guaranteeing the just rights of the oppressed majority and of the whole society.

In a democratic South Africa, major social programmes will be required to establish genuine equality between all citizens. The new South African nation will be built not simply on idealism, exhortations, and prayers but on the basis of progressively satisfying the material, cultural, and spiritual needs of the whole population in all spheres of life. The law has a key role to play in this process by establishing clear goals, appropriate institutions, suitable norms for each phase, and effective means of participation by all interested parties. The goals, means, norms, and institutions are not merely pious projections of what ought to be. They are given a firm statutory foundation, and appropriate legal responsibilities and corresponding legal rights are created.

The courts continue to play an important role as mechanisms which supervise enforcement and act as instances of last resort in the case of disputes. But they are not and should not be the immediate and principal agencies for the guaranteeing of rights. The vast social programmes that will be necessary in relation to pre-school and school-feeding, to the establishment of crèches and playgrounds, to mother and child health care, to mention a few key areas, require appropriate agencies with appropriate funding, functioning with appropriately trained personnel, according to appropriately defined criteria, and with an appropriate relationship to the community at large. If one looks at the field of health, for example, one may ask what is the more fundamental right — the right to primary health care, with an appropriate network of institutions and budgetary arrangements, or the right to sue your doctor for malpractice, with a corresponding network of lawyers and judges?

In relation to a charter of children's rights, one may envisage the law operating at six inter-connected levels.

Firstly, all laws which discriminate against children, forcing the majority to submit to segregation and inferior conditions, must be annulled. Race Classification, Group Areas, Bantu Education: every law, whatever its present name, that differentiates between children on the grounds of race, colour, or ethnicity must be scrapped.

Secondly, the instruments of the law — the army, the police, the prisons, and the judiciary — must be transformed so that they cease to be mechanisms of abuse and humiliation of children, and become means of genuinely protecting children's rights. Clear legal and disciplinary requirements must be established for punishing those in state positions who are responsible for crimes against children.

In the third place, the existing law relating to the protection of children must be strengthened and made effective in relation to the whole population. The specific rights children have in relation to their parents, namely the right to care and the right not to be abused, must not be lost sight of in the great general programmes that are necessary. Similarly, the right to a name and minimum inheritance should be guaranteed, bearing in mind different cultural traditions. The distinction between so-called legitimate and illegitimate children, so important in feudal-type societies, has to be eliminated as far as the general law is concerned. The role of the wider family, particularly relevant in traditional African society in relation to children, has to be given greater practical recognition, and community involvement in protecting children's rights has to be encouraged.

Fourthly, massive programmes of corrective action must be established to improve the condition of children's lives progressively and rapidly, guaranteeing minimum standards of nutrition, health care, housing, crèches, schools, sporting and cultural facilities. Such programmes will require a strong legislative foundation, appropriate budgetary backing, and clear mechanisms for planning and implementation involving government, public and private institutions and community bodies.

Fifthly, children's organization dedicated to representing the interests of children and securing their rights to live in peace, friendship, and equality, should have the support and protection of the law.

Sixthly, consideration should be given to creating a children's ombudsman post to handle questions, often of a delicate and controversial nature, related to the rights of the child. The children's ombudsman would be an independent figure with power to investigate cases and make appropriate recommendations. He or she would not be an alternative to the institutions of corrective anti-apartheid action, but complement them, acting on individual cases or localized pockets of abuse rather than on the broader level.

Finally, an appropriate textual mode should be found for proclaiming children's rights in broad and encompassing language. This could take the form of a declaration of children's rights which would act both as a standard-setting document and as a guide to the interpretation of existing law. Such a declaration or charter of children's rights would enter into South African national life and become a major point of reference and support for those struggling to ensure that new generations grow up in the secure and spontaneous conviction that South Africa, the whole of South Africa, belongs to all who live in it, and especially to all its children.

8 The future of South African law

Law teachers throughout the world seem to feel that, allowing for a few deviations and impurities, the system they teach is somehow uniquely good — the Americans because they have a constitution, the English because they do not have a constitution, the Portuguese because they have Codes, the Mozambicans because their system was revolutionary and had Codes, and South Africans because they had neither a constitution nor Codes nor was it revolutionary nor English.

Even professors who accept that the whole of public law in South Africa has been tainted by apartheid, argue that our private law is somehow remarkably good and that no one can sue for divorce, or breach of contract or damages for injury or for a right of way over land, with such confidence of justice being done as a South African.

We South African law teachers should not be so sanguine. In years past, we used to be required in our final year at the University of Cape Town Law School to do a paper called 'General Topics', which always included a question on the future of Roman Dutch Law (RDL). Although we were theoretically free to offer our opinions as we saw fit, we knew that the answer which would mark us out as intelligent and worthy of a law degree was that which said that RDL was uniquely suited to solve disputes, defend freedom, and promote commerce; in relation to the question of codification of RDL, although there were indeed weighty arguments in favour of such a course, in reality there was just as much case law over the interpretation of the Codes as there was in the common law system, so we might as well stick to what we had. The bolder amongst the students knew that we might tentatively suggest that the day was almost approaching when, in recognition of the contribution that English law had made to public law, commercial law, and the law of procedure, as well as in acknowledgement of the rich inputs to a South African jurisprudence made by generations of South African judges and law teachers, we could just possibly consider using the term South African Law.

The question now, however, is not what the law teachers on the examination board feel, but what the average South African citizen feels, and since there is no such thing in our divided society as an average citizen, what do the men or women who ride on the Putco omnibus or walk to work or, for that matter, are chauffeur-driven, feel on the subject? As someone who seems more and more to be going to work by aeroplane, I will offer my own views.

Every country needs a system of laws. It would be nice to think that in a democratic South Africa people will no longer die, but they will. There will also be robbery and theft and assaults; people will buy and sell motor cars and fridges and be injured by falling bricks. Post-apartheid does not mean post-dispute. There will have to be principles and procedures for dealing with disputes. Indeed, one of the main features of post-apartheid society will be the replacement of rule by arbitrary dictate — the essence of colonial and racist systems — by the Rule of Law. The democracy which we envisage is one in which the courts will be more rather than less available to the people, in which law and legal rules will have a greater rather than a lesser role to play.

In the case of South Africa, the rules that govern purchase and sale and insurance and companies and cheques, that deal with self-defence and dishonesty and when people should be held responsible for injuries to others, happen to have come from Europe. Like the railways and trousers and dresses and Bibles and the English language, they came in the context of dispossession and domination, but like the railways and trousers and dresses and the Bible and the English language, they have been taken over in varying degrees by the whole population and have now become South Africanized. Thus, after two hundred years, English has become a South African language: where else in the world can one say — he slipped on his guava? Shorn of their associations with domination, there is no reason why these institutions should not be taken over and infused with a new spirit so as to serve the people as a whole rather than just a minority.

It was the Verwoerdian ideal, not the popular one, to keep the mass of the population out of what he called the green pastures, to trap them in a distorted and frozen vision of the past, to exclude any dynamic for progress and transformation. This view was a continuation of the position of the Law Society in the Transvaal at the turn of the century when it opposed the admission to its ranks of Alfred

Mangena, advancing the argument that he was a native who should resolve problems through the traditional law and not through the courts. When Mandela made his famous denunciation of South African justice at his first trial after his capture, he did so with an elegance that enriched the patrimony of English usage in South Africa and, utilizing the principles and procedures of South African law to the full, he turned RDL into a weapon of attack. His basic critique against the legal system was not that it was Roman Dutch but that it was racist. Thus, he did not object to having courts with trained judges, to written laws, to defence and prosecution lawyers doing battle with each other according to defined procedures, but to the fact that he felt he was a black person in a white person's court; the laws were made by whites and administered by whites in a court-room that breathed the atmosphere of white domination, and this should not be so. He should be a South African in a South African court, he declared, and not someone subjected to a system whereby the guilty dragged the innocent before them.

One of the few advantages of exile is that it gives people the chance to work under and examine many different kinds of legal system. What appears to be increasingly evident is the growth of a universal legal culture, varying very much from country to country in the specifics of outer form, but increasingly developing a similar essence based on widely held principles of justice and of practical convenience. Many factors are involved: the existence of the United Nations and other international bodies, including the International Court of Justice, (to which one notes with pride, African jurists have made a notable contribution), the development of a global economy, the internationalization of communications and entertainment, even mass tourism. Although this legal culture is most strikingly expressed in relation to the universalization of principles of human rights and mechanisms for their protection, it also applies in less spectacular form to questions such as shipping law, international contracts, and air law, where similar language is used to achieve similar results, and commercial law, contract law, and even family law, where different language is used but the results are increasingly similar. Thus, the legal systems of countries as geographically and politically diverse as Zimbabwe, Senegal, Brazil, Cuba, Italy, the Soviet Union, Turkey, and the Philippines, all draw directly or indirectly on the principles of Roman Law, which should not cause surprise since Roman Law itself was the product of a process of universalization, drawing heavily on experience in Africa, Asia, and Europe.

Even if, for political or other reasons, it were one day felt necessary to start off the legal system of a non-racial, democratic South Africa with a clean slate, we would still have to have recourse to these universal principles. We might juggle them around a bit and give them a new packaging, but the results would be substantially similar to what we have today, save that they will have been received in conditions of equality and full sovereignty. On the other hand, there would be nothing to prevent the people, in the expression of their newly won sovereignty, from saying that rather than dedicate a vast amount of energy to re-inventing the wheel, they would accept the legal instruments at hand and develop them as much as possible in the interests of justice for all. In other words, RDL is already here, we know it more or less, we have the books and the rules and the procedures available, we might as well use it.

At this point a glance at post-independence experience in neighbouring countries should be instructive. The radical Independence Constitution of Mozambique in 1975 expressly declared that all existing laws that were not inconsistent with the Constitution remained in force. Thus the five Codes which contained the basic substance of the law, namely those dealing with civil law, civil procedure, criminal law, criminal procedure, and commercial law, all remained in force, and continue to this day to be the meat and drink of the courts from the district level upwards. What changed immediately was almost the whole of public law and certain aspects of family law. For the rest, the courts were re-structured, procedures were made less technical, and legislation was introduced over the years affecting certain economic activities and various aspects of criminal law. Zimbabwe, benefiting in part from counsel offered by the Mozambican leadership, gave even more emphasis to the importance of smooth rather than abrupt transformation. What one can say about both countries is that in the immediate post-independence period there are so many urgent legislative tasks to attend to relating to the very life of the new nation, that re-formulating the law dealing with such matters as murder or theft or liability for breach of contract or the issuing of shares or falsifying a cheque, comes very low down on the agenda. Tasks there will be in abundance for lawyers eager to make the legal system a vigorous and secure component of a non-racial democracy. Scrapping the common law and embarking on a vast programme of elaborating new codes is not one of them.

Does this analysis mean that a student writing a final LL B exam today, should give the same complacently conservative answer that guaranteed success three decades ago? Most certainly not.

In the first place, if what is referred to as RDL today is to survive, it must cease to be called RDL. The term Roman Dutch Law once had a faintly patriotic aspect inasmuch as it emphasized local particularity rather than Imperial connection. Amongst the great pioneers of RDL in South Africa were persons like Rose-Innes, Solomon, and Maasdorp who were neither anti-African nor anti-Boer nor anti-English. Ironically, the prestige they gave to the system was almost entirely destroyed by a more recent generation of judges who, under the guise of purifying and saving RDL, managed to combine extremely authoritarian views supportive of the apartheid state with a medieval scholarship so exquisitely recondite and impractical as to appear a parody of serious legal research. Thus if RDL is to survive, it must be purified of its purifications, and proclaim itself for what it is, South African Law. The term RDL is both inaccurate and insensitive, inaccurate in so far as it focuses on only one of the many sources of South African law, and insensitive inasmuch as it highlights the system's colonial origins. What is needed is a self-consciously South African law for an emerging South African nation.

Secondly, there is much in the legal doctrine itself that has proved to be out of date and oppressive: the dictatorial marital power of husbands in marriage, the feudal principles governing what is still referred to as the master and servants law, the wide scope of capital punishment, to mention just a few areas where legislation has not completely taken authoritarian principles out of the law. Land law needs to be completely revised so as to provide for flexible rules permitting concurrent interests in agricultural land, and inflexible rules preventing abuse of the land (*ius abutendi*). Contract law is still grossly out of touch with the rights of consumers, while the law of delict has to be completely re-thought in the light of new concepts of disaster insurance and no-fault liability. Rules of criminal law that permit lethal traps to be set against burglars and enable a security guard to kill up to twenty people with impunity, clearly have no place in the new South Africa.

Thirdly, the law is diffuse and not easily accessible. If progressive codification is considered too cumbersome and time-consuming a process at this stage, at least an authoritative re-statement of the law in compact and readable form should be undertaken.

Yet far more important than the alteration of name or even the adaptation of legal doctrine is the transformation of the judiciary and the legal profession themselves. Justice can neither be done nor be seen to be done until the judges and lawyers are far more representative of the community as a whole, far more sensitive in their functioning, and far more accessible to the population at large.

This means much more than appointing a few blacks to the bench. The courts must be truly South Africanized, so that everyone feels comfortable in them, with the result that persons on trial feel they are being judged by their peers and not by their masters.

In a period of transition it is particularly important that the judiciary enjoy the greatest respect of the widest sections of the community. It is not a question of lowering standards of procedure, decorum, or respect for the law, but of ensuring that the talents, probity, erudition, and experience of all South Africans are represented among the judges and among the court officials at all levels. Consideration should also be given to extending the assessor system and possibly reintroducing juries or other forms of lay participation so as to ensure a correct balance between professionalism and community involvement. What would be particularly damaging would be for an almost exclusively ageing white male judiciary, born and bred in apartheid South Africa, to have the final word on the great social processes required to bring about the end of apartheid (and, one hopes, of gender oppression).

How ironic it would be if judges, who for decades had obediently and slavishly implemented the most cruel of laws, were suddenly to take notice of John Dugard's criticism of legal positivism and begin to respond to his appeal for judicial activism and respect for natural law principles, just when Parliament becomes democratic and legislation is being aimed at enlarging rather than restricting the area of human freedom. South Africa has enough problems as it is without one day being saddled with a judiciary like the nine old men of the US Supreme Court of Roosevelt's time who declared unconstitutional New Deal legislation requiring employers to provide minimum conditions for workers, because it violated freedom of contract. In fact, bearing in mind the extreme delicacy and importance of interpreting the new democratic constitution, and the necessity of the judges involved enjoying the confidence of the whole community, serious thought should be given to the creation of a constitutional court consisting of persons of manifest integrity and wisdom, not all of whom need to have been practising lawyers (cf. the Italian, the Portuguese, and the Federal Republic of Germany Constitutional Courts).

Serious attention would have to be given to the way in which the provision of legal services can be opened up. Indeed, the legal profession itself should be opened up. Extensive programmes of affirmative action, preferably of a voluntary kind, will be needed to correct the enormous imbalance whereby less than ten per cent of legal practitioners are black, where blacks constitute nearly eighty-five per cent of the total population. The first thing to do would be to remove the obstacles to entry, which are not only financial and psychological, but practical: the need to know Latin, which is taught in white schools and not black, the difficulty of getting articles, not to speak of offices and chambers.

Legal training involves much more than simply mastering legal doctrine and procedure, important though this is. It requires knowing how to organize an office, manage money, deal with clients and colleagues and public officials, when to fight and when to settle. It involves questions of ethics, etiquette, and style, a vast semi-hidden body of knowledge that can only be acquired by practical experience. It is this area that appears to be most impenetrable and needs most urgently to be tackled. One realizes that legal offices have their own atmosphere and that a certain degree of intimacy and personal getting-on-together plays a major role in giving each firm its particular ethos; one is aware that clumsy and highly bureaucratic attempts to enforce change might result in a surface equality in which the real mastery of the profession continues to be in the hands of the few; yet these considerations cannot be an excuse for allowing legal offices to be self-perpetuating racist enclaves.

Given a serious will to tackle the problem, there is no intrinsic barrier to reconciling the wish of the existing practitioners to maintain what they regard as standards, with the right of all persons of suitable talent and integrity to earn a decent living in an interesting profession and the right of the community to feel itself represented in, and to have easy access to, an important social institution. Many factors could play a role: strong leadership within the profession, tax and other incentives for promoting equality, backing by local authorities, banks, and insurance companies for firms that provide equal opportunity and take active steps to encourage non-racism, and sensitively worked-out programmes of positive action supervised by the judiciary or other appropriate body.

Yet the problem is not simply one of access of would-be lawyers to the profession, but of access of the public to the legal system as a whole. All legal systems favour access by some sections of society over others, but few are as skewed as South Africa's. Apartheid means that those who are most dispossessed, whose rights are most frequently and flagrantly violated, whose legal problems are most dramatically severe, often literally involving questions of life and death, have the most tenuous access to the law. Yet those who already have power of every sort, economic, physical, political, not to speak of the immense confidence or arrogance that overlordship gives, can with a phone call or a command to a secretary or a nod and a wink at the golf course, summon up a battery of legal experts to do their bidding. A mixed economy with a vigorous private sector pre-supposes the mixed provision of legal services with a vital and self-confident private sector. In any event, to ensure that any future Bill of Rights operates effectively, it will be necessary to have a non-bureaucratized and spirited legal profession, unafraid to challenge arbitrary or unconstitutional action, whatsoever its source. But this does not mean that the virtual absence of community legal services, public defenders, and public interest legal groups can be justified.

At the moment, a handful of practitioners, some in private practice, others in legal resources centres, others at universities, are manfully and womanfully bearing an enormous burden. They operate on the margins of the legal mainstream, always overworked, frequently underpaid, handling the most challenging of cases often in the most difficult of atmospheres. The fact that they include some of the best and most imaginative legal brains in the country may be pleasing but does not eliminate the need to create conditions for the vast expansion of their number and the extensive opening up of the range of their activities. At the moment little more can be done than take test cases or cases with a relatively high-profile civil rights or political character. The vast number of day-to-day problems, frequently created or aggravated by apartheid, just cannot be attended to. One envisages the creation of neighbourhood and workplace legal services centres, financed by contributions from workers, employers, trade unions, local authorities, central government, and (dare one say it) from the private practitioners, either in terms of a direct levy or of some kind of turnover tax, for example on large conveyances or on the liquidation of huge estates after death.

The question of standards and traditions needs to be faced. More does not necessarily mean better, nor does it automatically lead to worse. In the South African context, more, if it means more representative, is a virtue in itself. What needs to be guaranteed with 'more' (without either practising racism or descending to paternalism) is the preservation and enrichment of all that is good for the development of the country.

Unfortunately, one cannot say that all the many long traditions of the profession are meritorious. It was the Cape Law Society that used the courts to stop women from practising law, arguing that because they were women they did not fall under the term 'person' in the relevant statute of admission. As has been mentioned, the Transvaal Law Society sought to prevent Africans from joining its ranks, and, more recently, tried to have Nelson Mandela struck off the roll because of his part in the Defiance of Unjust Laws Campaign. The Johannesburg Bar lost no time in having Bram Fischer, probably its most distinguished member of all time, struck off. In his autobiography, M. K. Gandhi speaks with emotion about the disdainful treatment meted out to him by his Natal legal colleagues. Until quite recently it was regarded as normal that black lawyers arguing an appeal in the highest court in the land at Bloemfontein should go to their cars and drink tea from a thermos flask while their white colleagues had refreshment offered to them in the court building. It was part of the tradition that black advocates should robe up in separate rooms, that black lawyers should suffer the indignities and disadvantages imposed upon them by restrictions on occupying premises in so-called white areas, and that until quite recently, white witnesses should be referred to as mister or miss and blacks by their first names. This profoundly racist tradition is still there, even if in less flagrant forms.

Then, at another level, far from being traditionally fearless, the organized profession has shown itself to be notably fearful, or even worse, indifferent in the face of repeated invasions by the legislature and the executive of basic rights and liberties relevant to due process — one thinks of areas where legal and judicial functioning are directly affected, such as declarations of states of emergency which go on for years, detention without trial, the bringing of witnesses to court straight from months of solitary confinement, the denial of access of detainees to lawyers, and the direct or indirect ouster of judicial review.

Happily, there are other traditions to which one can point with pride. We do not have to have recourse to a Grotius or a Coke to find legal freedom fighters in our past. Gandhi, Schreiner, Krause, Seme, Mathews, Fischer, Nokwe, Berrange, Kahn, Muller, Mandela, Tambo, Slovo, and Kies, the list is long and can be made much longer: persons drawn from every section of our community who saw the pursuit of their legal careers as being inextricably linked up with the pursuit of justice. The list is even longer of those who, without confronting the system of injustice head-on, have used their legal talents to defend those dragged before the courts under apartheid or security laws — Pitje, Kentridge, Bizos, Mohamed, Kuny, Aaron, Cheadle, de Villiers, Richman ... the list is being added to by the day. Perhaps of special interest is the role played by certain far-sighted and fair-minded judges over the years. Rose-Innes was a judge of whom any country could be proud, and he is as much part of our patrimony as the hanging judges of today. An outstanding legal scholar who researched and adapted RDL to modern conditions, he imbued his judgements with as much of the spirit of liberty and equality as he could. Krause, a courageous Boer freedom fighter, imprisoned for secretly belonging to the MK of his day, was a life-long opponent of capital punishment and fighter for penal reform. There are judges on the Bench today unafraid to stand up to the securocrats and to voice their judicial distaste of cruel and oppressive edicts; it is not necessary to identify them, their open-minded judgements proclaim who they are.

Open-mindedness, tolerance, a sense of basic justice, respect for the essential dignity and equality of all ... these qualities have never been absent both on and off the Bench, but always the system of racial domination has inhibited their expression. The creation of a non-racial, democratic South Africa will for the first time liberate these qualities so that they become the norm and not the exception in South African legal life.

There are also other qualities which can be retained and developed in the post-apartheid legal profession. Mixed in with the racism and indifference to suffering are certain ethical and professional standards that are eminently worthy of retention, such as never consciously misleading the court, fair dealing with clients, keeping one's word with colleagues, attention to meticulous book-keeping, careful preparation of cases, elegant presentation of documents, and an overall sense of professional honour. South African lawyers have a reputation for conducting themselves with spirit and yet with decorum, and it would be a sad impoverishment for the whole country if these characteristics were lost.

In the period of transition from a closed, apartheid society to an open, democratic one, imaginative and vigorous lawyers will be required to ensure that creative solutions are found to the difficult problems of reconciling conflicting interests in a fair way, and also to ensure that the bad old authoritarian habits of the past do not slip back. Freedom will not have been brought about by legal toadies, nor will it be preserved by such. Better an honest and straightforward reactionary who does his or her work in a straightforward way than opportunist and corrupt lawyers who undermine the spirit of true equality by paying lip-service to the new while really believing in the old.

Yet even changing the name of RDL, opening up the courts and the legal profession, developing alternative forms of access to the courts, and drawing on all the positive aspects of legal tradition in our country, will not be enough. These steps will make the law more open, sensitive, and democratic, and enable the legal system to play an active role in promoting orderly change while respecting the rights of individuals; they will help to guarantee the tranquility the country will need and which the mass of the people have not known since the first days of conquest and dispossession. But they will not in themselves take account of the cultural and political fact that what we sometimes loosely refer to as the legal system, in fact draws on three and not just two sources of law, namely RDL, English law, and African law. In reality, the full dimension of the fact that South Africa is, after all, an African country peopled by persons of diverse origin, has never been addressed. Far from being a fundamental part of the legal system, the indigenously African component has been relegated to the margins of RDL, at best tolerated and barely recognized as a kind of 'own affairs' activity taking place under white tutelage, at worst manipulated to justify the bantustanization of the country.

It is unthinkable that in a democratic South Africa this neglectful and insulting attitude can continue. It is not simply as a matter of respect that traditional African law, both in its earlier and in its contemporary forms, needs to be properly studied and understood, but that African tradition contains many elements and resources that could enrich and invigorate the whole legal system.

Many hard questions will have to be asked. One of them is whether to have a unitary system of law in terms of which the same rights and duties attach to all South Africans, independently of cultural background, or whether to admit a degree of pluralism whereby there could be different rules for different sections of the population in certain areas such as family law, succession, and rights to land use. Such pluralism could be on a personal law basis, that is, apply to persons of a certain cultural background anywhere in the country, or it could be regional, or based on choice by the parties concerned, or on a combination of any or all of these criteria.

Many variants are possible: there could be a unitary court system applying pluralist solutions depending on the background and, perhaps, the wishes of the parties, or there could be separate court systems with the judges themselves coming from different backgrounds and using different procedures, or a pluralist system at the base supervised by a single higher judicial echelon. Put more concretely, there could be a national, non-racial system of judges and magistrates applying a single national law. Alternatively, the same single national system could operate but permit, say, traditional African legal rules to be applied by the judges or magistrates in a particular case. Another possibility is to create or recognize special legal institutions based, say, on traditional leaders or imams, with competence to handle cases concerning marriage and divorce of persons with a certain cultural background. Finally, the decisions of the traditional leaders and imams could be made reviewable according to certain basic constitutional principles by the national courts.

In the past, these sorts of questions have been determined in the context of how best to serve colonial or racial domination. The persons most directly concerned have had the least say in finding the answer. The big difference in the future will be that everyone will participate in the debate, and that special weight will be given to those for whom the issue has the greatest practical and sentimental meaning. Thus, this is eminently not the sort of issue on which one should try to pronounce with confidence in advance.

My own intuition, and it is nothing more than that, is that the people will increasingly opt for a single national system in which the basic equality of rights and duties of all South Africans, independent of language, colour, ethnic background, or religion, is emphasized. Apartheid has given pluralism a bad name, and the yearning for unity, particularly in relation to public institutions such as the courts, is powerful. At the same time, the people will be free to celebrate and dissolve unions according to tradition, and apply the traditional rules if they so choose. This already happens in the case, say, of Catholics, who are free on a voluntary basis and as a matter of faith, to apply the rules of canon law to their marriage (for example, the prohibition on divorce) but who as citizens may, if they so choose, go to court and invoke the principles of the state law (which do permit divorce).

Similarly, for many Jews, the social and religious consequences of their marriage in a synagogue are in practice far more important than the purely legal effects as determined by state law; nevertheless, if a case involving a Jewish couple comes before the courts, the rules of rabbinical law are disregarded, and the general law of South Africa is applied.

In the case of traditional African marriages, this would mean that the question of whether or not lobola had been paid would be one for the families alone and not for the courts. Criteria could be established for recognizing traditional African marriages, as well as for attaching rights and responsibilities to persons who enter into serious relationships without the formality of marriage. But once recognized, certain basic principles of family law could apply to these marriages or unions, independently of the rites followed. As a matter of social custom, the families could continue to argue over lobola, but not as a matter of law.

My sense is that people would prefer a system which was Africanized in a number of ways but not in such a way as to encourage ethnic division and endless disputes over intricate matters of family relationships. Thus they would want to see Africans on the bench, and to hear their problems discussed in the languages with which they are most at home. They would like the procedures to be more flexible, courteous, and open, in the way of traditional African courts, and for certain generalized aspects of African culture to be recognized by the law, such as consultation with the family council where possible before marriage and before divorce, and strong attempts at reconciliation by discussion and mediation.

In particular, there is one immense legal resource which the present system is incapable of utilizing that could play a vital role, particularly in family and neighbourhood disputes. There already exist in South Africa a vast number of informal courts that operate amongst the people wherever they live and work. These courts have little in the way of written rules, but they are used extensively to solve acute problems at the local level. Clearly they draw heavily on African tradition, and indeed, testify far more to the vigour and adaptability of traditional African law than do the textbooks on what used to be called customary law. The challenge facing us will accordingly be how to take advantage of this immense community involvement in dispute resolution, in a way that neither crushes its spontaneity on the one hand nor permits manifest injustice on the other. The community courts envisaged here should not be equated with the so-called *makgotla* or with the people's courts which were created or sprang up in various parts of the country in recent years. These types of courts operated in conditions of state repression and local insurrection. They manifested the weaknesses of the existing legal system without necessarily pointing to answers. We need to study their experience closely, not on the basis of which side any particular court supported, but of how it worked and what its relationship with the people was. It is obvious that the courts functioned in many different ways, some being excellent by any standards, others disastrous by any criteria. Community justice is not in itself more fair or less fair than any other form of justice. Its ambit needs to be clearly defined, its competence limited. While serving the positive function of involving the community in the settlement of disputes, it should never serve the negative purpose of undermining people's constitutional rights, and especially never diminish every citizen's right to a fair trial. The community wants the security of knowing that its members have guaranteed legal rights. There should in principle be no contradiction between community justice and the rights of due process. The problem is to combine the two.

On the basis of a decade of experience in Mozambique, a country which has bled a lot and seen many of its hopes crushed, but which has also scored important successes in certain fields, the handling by the community courts of family matters being one of them, I would say that community courts would have a very strong future in South Africa. They would draw heavily on African tradition in that they would function in a less formalistic and professionalized way than the existing state courts, they would look at questions in a multi-faceted rather than purely technical manner, and they would be made up of several members who would try to reach their decision by consensus. They would transform and modernize African tradition, or rather, reflect the new African tradition that incorporates trade unions and church groups and community organizations, through the inclusion of women and men, young and old among the judges, and by applying practical, common sense, and manifestly just solutions to the concrete problems before them. Thus in the case of family breakdown, they would be more interested in the assets of the parties, the question of the home and how the children can best be protected, than in pursuing all the ins and outs of the pre-marriage negotiations. The courts would operate at the grass-roots level only, and not have the power to deprive people of their liberty or impose corporal punishment.

These intuitions could turn out to be completely wrong. The crucial thing is that the people will be consulted and directly involved at every stage. The role of lawyers will be fundamental in terms of helping to frame concrete questions to be put to people, and in finding appropriate institutions to materialize people's wishes.

Thus Roman Dutch Law, which survived in the past by incorporating transplants of English law, will survive in the future by fusing itself with African law, shedding its name, and becoming an integral part of a new South African law.

9 Rights to the land

A fresh look at the property question

The one thing that does not grow on land is land. However one looks at it, the surface area of South Africa is limited and not even the advent of non-racial democracy will make it larger. You cannot extend land rights in the way you can extend the vote. The land is not only finite, it is fixed; there is no way of physically redistributing and re-locating it the way you can with money or cattle or bags of maize. The land is the land. You can fly over it, tunnel under it, wash its surface away, put up buildings on it, degrade it, beautify it, live on it, abandon it, and in the end it is just as big or just as small as it was in the beginning.

At first sight the land question seems to be yet another of South Africa's many allegedly insoluble problems, perhaps the most difficult of all. Either the original unjust dispossession of the land is condoned and recognized as a legal fact, or there is a new form of dispossession

---

I know a little bit about the law. Aninka Claassens knows a little bit about the land. Between the two of us we tried to develop an approach that seemed consistent with the values and wishes of those on the land as well as with the general movement towards democracy and nation-building in South Africa. A third author, although he was not aware of his collaboration, was my friend and colleague, Dr Alpheus Manghezi, who over many years has helped me and others to see the importance of listening to the people on the land and respecting the common humanity to be found in the cultural values of all, both black and white.

I take responsibility for the presentation of the argument. All the themes need following up and further thought. The factual foundations require rigorous investigation. Nothing stated above should be regarded as representing official or unofficial positions of any organization. The objective was merely to get some new ideas into circulation and to enrich the ongoing debate in terms of which people of the land must be principal actors.

which, it is said, would unjustly deprive the present owners of what they have legally bought or inherited and developed with their money and their sweat. What would be transferred would not be land but resentment, and the only issue would be who should bear the anger: the original possessors, currently dispossessed, or the current possessors, about to be dispossessed.

In any event, quite independently of the justice argument there is the food argument. The whole country needs food — the reasoning goes — and any major re-allocation of access to land, particularly if it involves replacing skilled by unskilled farmers, would so undermine agricultural productivity as to ensure that the only equality that South Africa would get would be the equality of hunger; the whole process of consolidating democracy would be jeopardized, and black farmers would suffer like the rest. Experiments in new forms of land ownership, such as state farms, collectives, or co-operatives, would simply add to the confusion and hasten the collapse. The problem can be mitigated, the argument continues, by a massive injection of money, and by looking for unused or abandoned land — but it cannot be solved. The corollary of this proposition is that iniquitous though the present division of land might be, it is better not to interfere too drastically: rather the disaster we know, which we can blame on history, than the disaster we do not, which will be attributed to us.

South African land has not always produced food, but it has always been fertile ground for producing questions.

The terrible statistic eighty-seven versus thirteen (representing respectively the percentages of land reserved for white and black ownership and occupation), created and endured by our ancestors and lived by ourselves, cannot be avoided. Our past weighs on us like the Drakensberg. The question is thus one of how to alter these proportions, so that the ownership ratios correspond more directly and not inversely to the population figures. This means that just as land was taken from blacks because they were black, so in future must land be taken from whites because they are white. The issue is whether this should be done suddenly or in stages, with or without consent. The second question revolves around compensation: how, it is asked, are whites to be recompensed for giving up their rights in land, who is to pay, and with what? The third question is posed in the form of asking what type of economic and legal regime should be adopted in relation to re-distributed land. Should large-scale farming be maintained in the form of state-owned or co-operative farms, or should land be parcelled out to small-scale family farmers?

It is suggested that formulating the questions in these ways makes a solution more difficult than it need be. They are cast in a generalized and abstract manner, whereas land is very concrete, and people's claims to it are quite specific.

These questions encourage international searching for models, whether of success or of failure, to fuel arguments rather than the involvement of the people most directly affected in discovering answers. They overplay the commandist aspect of working out solutions and underplay the potential key to the whole issue, namely, the wishes and culture of the people already on the land.

Putting the questions like this runs the risk of appearing bold and favouring the dispossessed, but actually ending up timid and supportive of the *status quo* because ultimately the questions seem insoluble. What follows is an attempt to lay the emphasis on principles and procedures rather than outcomes, and to situate then land question in the context of democracy, human rights, and the Rule of Law rather than the context of race.

The sovereignty dimension

To this day, the fundamental question in relation to land is that of sovereignty and de-racialization. As long as race is the determining factor in deciding ownership and control over land, every struggle over every square metre will be a racial struggle. Only if we truly de-racialize the terms of ownership, occupation, and use, will the question really become a question of land and cease to be a question of domination and subjugation.

South Africa has been appropriated by a minority. At the political level this appropriation has been maintained by monopolizing the franchise, at the level of daily life by control of the land. The fact is that whites by law own eighty-seven per cent of the surface area of South Africa. They can expel blacks from the land, demolish their homes, prevent them from crossing or remaining on the land. Control over land is not only control over a productive resource, it is control over the lives of people.

The racialization of land ownership began with the first wars of conquest and continued with appropriation through treaties and direct occupation. The dispossessed African population tried to retrieve their land by purchase; they were forbidden by law, as 'natives', from doing so. They then sought to retain access to the land as lease-holders; they were prevented by law, because they were black, from doing so. They entered into agreements as share-croppers; these agreements were invalidated on grounds of race by law. They worked the land as labour tenants; this was made illegal in terms of so-called native policy. Those who had managed to cling to legal title were forced to vacate their land because they were said to be occupying black spots in white land. Millions of persons had their homes bulldozed, were carted away in lorries, were physically expelled if they ignored the legal notices that ordered them to remove themselves from so-called white areas. They moved back to the land. They were prosecuted as trespassers.

This is what Chief Albert Luthuli, President of the ANC, and member of a successful African sugar-farming co-operative, was referring to when he posed the fundamental question: who owns South Africa? He answered as follows: 'The overwhelming majority of whites, because they are "white", extend their possession to the ownership of [black] people, who are expected to regard themselves as fortunate to be allowed to live and breathe and work — in a white man's country.'

Furthermore, everything in relation to land utilization was organized on an explicitly racial basis: loans from the Land Bank, credit, marketing, the provision of services, subsidies, the extension of transport, the system of taxation, and exemptions. White farmers benefited even in relation to other whites. The franchise was loaded by twenty per cent in their favour, they were grossly over-represented in Parliament and able to influence legislation in their favour, down to such shameful details as compulsory flogging for stock-theft and the abolition of school meals for black children.

Two completely different and unequal systems of land law emerged, one for whites and another for blacks. Land law for whites was based on private property, registration of transactions in relation to land, ownership proved by certificate of title and demarcated plots. Land could be leased or used as security for loans by means of mortgages. The owner as property-owner was sovereign, a little king or queen over such land as was registered in his or her name. He or she could dispose of it at will, sell it, lease it, give it away, even control its destiny after death by means of a last will and testament. Subject only to planning permission, the owner could do what he or she wished with the land: use it, abuse it, dig holes in it, or do nothing with it, just own it.

Black land, on the other hand, was state-owned and controlled. Access to such land was governed by a system of grants, rigid laws of succession, and supervision by government-appointed or recognized chiefs. Occupiers could grow food there, erect houses, and, subject to controls, keep livestock on it.

What is clearly needed, if we are to resolve the issue of sovereignty and reach the real question of how the land should be owned and worked, is the nationalization of land law. For those who quake at the word nationalization, let it be stated firmly that nationalizing land law does not presuppose either nationalizing the land or nationalizing the legal profession, but simply ensuring that South Africa has a single, or national, law governing the question of land rights, so that issues are no longer addressed in terms of race, as at present, but in terms of interests and values of importance to the country as a whole.

This obviously requires the immediate abolition of the Land Act and the Group Areas Act which explicitly divide the surface area of South Africa on racial grounds, as well as the repeal of laws which permit forced removals and banishment of blacks. Yet it necessitates far more than that. Nationalization of land law means establishing in positive form an integrated, nation-wide legal framework in respect of interests in land. It presupposes South Africanizing the law, that is, having a law for South African citizens, whether they be farmers or householders or visitors or builders. The content of the law must be South African, that is, it must derive its principles from the values held in relation to property rights by all South Africans, embodying and being enriched by different cultural and legal inputs. In its formulation and application, the law can take account of different local situations — whether land is urban or rural or park; it can allow for different patterns of farming, or even of forms of tenure; it can respond to different claims of the people on the land, for property rights in some cases, workers' rights in others, and citizens' rights in all. What will go will be any reference to race, or any differential provision of services on the grounds of race, or any assumption that the whole of property law has to be fitted into the principles of Roman Dutch law.

Nationalizing land law will have immense implications for the relationship between the state and farmers. Instead of differentiating between white farmers and black farmers, the former to be helped, the latter to be controlled, state institutions will simply look at South African farmers, all of whom will have equal claims and entitlements in their capacity as farmers and not as whites or blacks. At the moment, there is no area of activity in which the unequal provision of services is more pronounced than in agriculture. One can say that there is massive affirmative action — in favour of whites. The first thing to do will be to end the vast privileges attached to race as such, and to ensure that what the state supports is farming and not whiteness. The question of subsequent affirmative action to support the racially underprivileged rather than the racially overprivileged will then be one that can be considered.

Yet something far more profound even than equal access to land and equal state support is necessary. We will have to tackle the way in which racially-based land law undermines the fundamental human rights of the citizen. At the moment, land law, instead of being a bastion of personal freedom and independence, serves as the basis of the most blatant denial of basic rights. Because control of land presently means control of people, white landowners exercise a double sovereignty in relation to land: they are kings and queens both in relation to what the law says is their territorial domain, and in respect of the people who are born within or enter that domain.

The only security that blacks on white-owned land have is the precarious goodwill of the landowner. However ancient the connection of black agriculturalists with the land might be, the law only has regard for the will and interests of the persons who own the title deed. The courts declare black farmers or householders to be squatters or trespassers. At best they have a right to a short notice period before being expelled. At worst, they can be imprisoned for being on the land against the owner's wishes. One is not referring here to casual passers-by or escaped criminals. One is thinking of people whose parents were born on the land, and their parents before them; people who have no right to be on any other land, who have no other home than the one they constructed themselves on the land from which they are being thrown out; people whose only wish is to have security and be able to earn a decent living.

In this setting of legal domination, there are few restraints on physical domination. White farmers feel free to decide who may visit black farm dwellers, frequently to demand casual services as of right, often to enter workers' homes without invitation, and sometimes to abuse them physically.

De-racializing the law and giving it a truly national character accordingly requires that the rights of persons in relation to land be integrated into and harmonized with a system of constitutional rights, and subjected to the principles of the Rule of Law. The hard legalism of the English common law to which Max Weber made reference, has to give way to humane concepts of rights as enshrined in a Bill of Rights. There has to be respect for the person, for the home, for freedom of movement, for secure family life, on the platteland as anywhere else. A person should be no less free because his or her home happens to be on spot B rather than spot A, or because Baas or Madam thinks he or she is well-behaved or cheeky. Equally, his or her rights to education or medical attention should not be qualified by whether a particular landowner is enlightened or backward.

Finally, nationalizing the law in the sense of making its rules cover the whole nation and not stop at the boundaries of this or that farm, presupposes the extension of the principles of legality or the Rule of Law over every square centimetre of the country. The police force and the courts should be there to defend equally the rights of everybody, and not serve, as overwhelmingly they do today, to impose the domination of landed whites over landless blacks.

Conclusion

Abolishing racist statutes, equalizing state supports, introducing principles of constitutional rights, and applying the Rule of Law are the concrete ways of de-racializing land law and opening the way to a fair and widely accepted method of tackling the difficult problem of competing claims to land.

De-racializing land law is not just bringing the Rule of Law to all aspects of rural society in a non-racial way, though it includes that. There has to be a de-racialization of land law as such, that is, of the law governing the control, occupation, and use of actual pieces of land. The whole of property law has been debased by racism. It is more than just a question of who can and who cannot be owners. The very meaning of property rights has become increasingly degraded. Increasingly, the rights have had less and less to do with the actual relationship of people to the land, and more and more to do with whiteness. Property law has ceased to be an instrument for protecting true property values, and become a means of preventing competition from black farmers and proletarianizing all Africans on the land. We have a law dealing with property in South Africa, but not a true property law.

It is ironic that those who over decades and centuries have converted land law into an instrument of racist domination, should now be the strongest defenders of what they call a neutral property law, by which they mean a law which will defend the existing ownership patterns as ownership and not as white privilege. It is ironic that no one has done more to undermine genuine national respect for property rights than the capitalists, whether on the farms or on the mines, and no one has worked harder for recognition of the property rights of the people than those who have regarded themselves as anti-capitalist.

Whatever one's philosophic starting point might be, the notion of property connotes a degree of legally guaranteed security, independence from arbitrary interference, the right to contract freely, and the order of an abiding and objectively determinable system of principles and procedures. These values, shared by farmers from the most varied backgrounds, have to be disinterred from the rubble of apartheid law, which has targeted each one of them for destruction. The new law requires more than just the absence of the old. It needs the reconstruction of the values to which the old paid lip-service but which it systematically denied.

Non-racism, against the background of racism, necessitates more than the existence of technically neutral rules governing the future acquisition and use of land. It presupposes drawing on the experience of all South Africans in relation to the land, listening to them all, discovering common points of resonance, and involving all in the processes of transformation. Solutions found in this way are likely to be more concrete and enduring than those thought up by think-tanks, however enlightened or progressive the experts might be. At the very least, all those most directly connected with the land should be given the chance to participate actively in the processes, so that those who are seriously committed to maintaining good farming, whatever their background, have the chance to make their contribution.

Only if these democratic principles are followed can the question of sovereignty be removed from the land question and the true societal values in relation to land common to all cultural groupings be uncovered.

This is what the Freedom Charter demanded when it said that South Africa belongs to all who lived in it, and that the land should be shared amongst those who work it. Once the principle of a common belonging is established, the basis of equitable sharing exists. Until the foundation of common belonging is laid, however, defence of private property means defence of white property, which means defence of white domination.

The generous and far-sighted statement which opens the Freedom Charter provides the foundation for a new land law in South Africa. Once it receives an echo from present property holders who are willing to see beyond race and acknowledge the commitment to the land of all those who work it and live by it, it becomes possible to build up a shared set of principles for the new land law and to agree on a new set of procedures for deciding on competing claims in relation to particular pieces of land.

The values dimension

As far as the majority of South Africans is concerned, present land law lacks legitimacy. The land was taken by force and deception. The structure of white domination, which subsequently registered title deeds and created a market in land, was itself illegitimate. To this day, race is the foundation of property rights, both substantively and technically. Any attorney's typist knows that the deeds must set out the race of the parties, otherwise they will be rejected by the Deeds Office. Racial compatibility is the foundation of legal efficacy.

To add to the sense of illegitimacy, blacks were forced by the pass laws and the system of what was called native taxation, to work on the white-owned farms. Illegitimacy tainted the use of the land as much as its acquisition.

Getting rid of the overt racism in the law and creating conditions where land is seen as land and not as power, is therefore the foundation of re-legitimizing land law. Yet it does not in itself indicate what the sources of a new land law will be. Simply to rely upon the existing title deeds as the basis of property interests in a new South Africa would be to evade the issue of legitimacy altogether instead of confronting it. It might appear to have the advantages of convenience, but in fact it would prove totally inconvenient, since it would guarantee that the sovereignty debate would continue on a plot by plot basis, even if formally resolved at the national level. In many parts of the country, such as in the South Eastern Transvaal, Northern Natal, and Northern Free State, black farmers are aware of dispossession not simply on a generalized historic basis but in relation to specific pieces of land farmed by themselves or by their ancestors. As far as they are concerned, the title deeds possessed by the present white owners have no legitimacy whatsoever, since the original titles from which they purport to derive their effectiveness were themselves tainted with illegitimacy.

It is true that in all countries present-day property relations are based on ancient acts of conquest or forms of internal appropriation which in contemporary terms would not bear legal scrutiny. No one in England would seriously seek to set aside present landholding arrangements because historically they emanated from a system of tenure introduced by William the Conqueror. In North America the land claims of native Americans are based on treaty rights in relation to specific areas of land rather than original possession of the whole continent. The difference in the case of South Africa is that blacks continue to occupy and till the soil everywhere; that the colonial character of the relationships between themselves and the whites who have legal title has been overtly maintained by the law and state practices; that a vast range of treaties, grants, and contracts existed to cover large portions of the country before they were subsequently unilaterally reneged upon or repudiated by the whites; that the black peasantry never forgot their original rights and never ceased to struggle to restore them.

A close look at the demands of the dispossessed shows that they take different forms in different parts of the country. In and around the bantustans and in the areas where black farmers have managed, despite all the attacks on them, to cling to their land, notably in the eastern parts of the country, the pressure is for rolling back the years of encroachment of neighbouring white farmers on their land. In areas like the Western Cape, the Western Transvaal, and the Natal Midlands, where proletarianization of the farmers goes back further and has been more profound, the claims at this stage appear to relate to securing dignified conditions of work and pay rather than getting direct access to land.

Whatever the position on the ground, two basic principles must be followed: people must be consulted and involved in any process of change, and the new property law that emerges must be based on a shared patrimony of values.

Private property, whatever its precise legal form, is said to be based upon certain social values of an enduring kind. These are:

Security

The owner has a stable and unbudgeable interest that will be recognized by the state and the whole world. It can only be interfered with in the limited circumstances where the law permits expropriation in the public interest subject to compensation. In the case of South Africa, the whole intent and thrust of property law has been to deny stability of access to and use of land on the part of blacks. Even to this day, the limited property rights blacks enjoy in the so-called black areas are simply ignored when black families are removed from other areas and dumped there as though on open land. What matters is blackness, not rights of possession.

Independence of the landowner from state interference

Private property implies an acknowledgement of domain. State functionaries need a judicially authorized warrant to enter a person's property, and once inside, are obliged to respect the physical integrity of the property. The state cannot tell the owner how he or she should use the property, save for imposing certain parameters of choice through planning and environmental controls. In the case of South Africa, the state has deliberately set out to undermine any notion of blacks having independence in relation to land. The state sends in its police and its bulldozers, its cattle-culling inspectors and its native affairs officials. It moves people from one area to the next, demolishes homes, forces reduction of cattle, decides who may or may not visit the land.

Freedom to contract in relation to the use of the property

In South Africa this right has been systematically denied. The principal objective of the Land Act was to prevent blacks from entering into contracts of sale or lease. Contracts which blacks have solemnly made with white landowners, such as share-cropping arrangements or agreements for labour tenancy — tenacious attempts under conditions of unequal bargaining power to establish continuing legal connection with the land — were later deliberately and directly undermined by successive apartheid statutes.

Stability irrespective of changes in government is currently being asserted as one of the hallmarks of private property. In South Africa, governments have come and gone specifically on programmes of reducing black property rights. Indeed, no question of black rights in land existed; all that Parliament recognized was what was called black policy and the administration of black people. Now that white-owned property appears to be under possible threat, the virtues of respecting vested property rights are being discovered. It is said that once the Land Act and the Group Areas Act are repealed, property law will have been de-racialized and it will then behove all true supporters of the system of free enterprise to defend existing property arrangements. Yet the reality is that enterprise has never been free in South Africa. For the majority of the people (black), it has been totally under (white) state control, totally regulated (by whites) and totally monopolized (by whites). Defenders of free enterprise should thus be the last persons to demand that present patterns of ownership be respected.

One thus sees that, point by point, the claims made in respect of the virtues of the system of private property have been controverted, deliberately one by one, by the very persons who allege that they are the true defenders of the system of private property. Conversely (and to complete the paradox) black farmers, who allegedly have no understanding of or interest in private property, have fought vigorously and against increasingly heavy odds to retain respect for these true values of proprietorship. Their decade-long struggles to recover their rights to lands from which they have been expelled, prove the depths of their attachment to the soil, not to any piece of soil, but to this or that plot that they regard as theirs by birthright, occupation, or contract. Any lawyer dealing with land claims becomes immediately aware of how deeply meaningful property rights are to black farmers, for whom the notion of property goes well beyond simply having a right that can be computed in money terms and becomes one of close relationship to and responsibility towards a particular piece of land. The land represents the link between the past and the future; ancestors lie buried there, children will be born there. Farming is more than just a productive activity, it is an act of culture, the centre of social existence, and the place where personal identity is forged.

Re-legitimizing the law accordingly requires that these values, proclaimed in theory but repudiated in practice by whites, be restored to their proper place, which is right at the heart of the concept of property. Security, independence, the binding nature of contracts, and continuity of rights — this is what the black farmers are demanding. The issue they are raising is not whether to have large traditionally organized communal farms or modern-type co-operatives or small family farms, but whether to acknowledge their concrete and usurped rights to property. Once that has been done, the people, in the exercise of their free choice, can decide whether to continue farming in family units or whether to merge, reintegrate, or divide their fields. The basic question on the agenda right now is one of legitimation, not of parcelization or collectivization.

Two important and interconnected consequences flow from placing the emphasis on values rather than on race, and from stressing legitimation rather than economic forms.

In the first place, such an approach avoids conceiving of redistribution simply as a racially quantitative procedure. If whites say today: we own nearly nine tenths of the land because we are white, and not because of certain values, then they ought not to be surprised if blacks go on to answer: we want three quarters of the land back because we are black and constitute three quarters of the population. Not only would a re-distribution conducted on such a mechanically mathematical basis keep the racial principle alive and guarantee sabotage by the present owners and total disruption of the food supply, it would do nothing to establish criteria for preferring this or that new claimant to a particular farm. Thus, if two claimants satisfied the qualification of being black, the land would go either to the one who had the most money or to the one most favoured in terms of influence or else best placed in a bureaucratically organized queue; in either event, gross injustice could result, and persons with ancient connections to the land could be shut out. New property rights would flow from new title deeds, not title deeds from intrinsic property rights.

Secondly, the proposed values-based perspective provides scope for shared, legally-protected interests between black and white claimants to the same piece of land. Where shared values exist and a shared commitment to and involvement with a particular piece of land exists, there is no reason in principle why the law should not be adapted so as to cater for and protect such shared interests. Sharecropping and labour tenancy in the past were examples of co-involvement between black and white in production on a single farm. The arrangements were based upon contracts acknowledging the fact that black and white families occupied and farmed the same piece of land, and defining the mutual rights and responsibilities between them. In the conditions of the time, the parties contracted on a grossly unequal basis, in terms of which the white farmer was accorded a dominant position and the black farmer a subordinate one. What will become possible in the period of democratic transformation in which the human rights of all are acknowledged by the constitution, is a re-negotiation of the terms of shared occupancy and use, but this time on the basis of objectively determinable criteria and in an atmosphere of equality. Negotiated contracts involving the people most directly concerned have the advantages of encouraging solutions which take into account concrete realities, including the preferences of the parties themselves. Such contracts would be far more likely to become operational than determinations imposed from outside.

Respect for shared values could be the foundation for solving many of the acute problems of reconciling competing claims to the land. Yet the question is not simply one of rights to the land, but of rights on the land. What all farmworkers are demanding, whether they are peasants seeking to get their land back, or rural workers trying to improve their conditions, is that their human rights as people be recognized. Without integrating the values dimension into a system of generalized respect for human rights, it has little chance of being meaningful in South African life.

The human rights dimension

The question of property as a human right has been turned inside out in South Africa. The issue is presented as though the one fundamental human right in relation to property is the right not to have your title deed impugned. All other aspects, your right to a home, to security, to independence, are ignored if you do not possess the title deed.

Your actual relationship with the land is totally irrelevant; you buy the land, you buy the labourers. You might be living on a Greek island, you might have bought the farm because you are making so much money from other activities that you need an investment which guarantees an income loss and a hefty tax rebate, you might have acquired the farm with taxpayers' money in the form of a massive low-interest loan from the Land Bank which you never pay back, you might be making a reasonable income not because you are a good farmer but because you are white and entitled to subsidies and guaranteed prices. Yet according to this approach, any intervention in respect of your relationship to the land would be a gross violation of your human rights.

Conversely, you might be a descendant of generations of persons who have lived on and farmed the land in question. You might have been born there, regarded the land as intrinsically the land of your foreparents, invested your sweat and thought into the soil during years of drought and years of plenty, brought up your children there. It might have been your one and only home since birth, your only workplace, your one place of social life. Yet the argument would be that you are nothing more than a squatter infringing the rights of the true owner of the land.

The basic fact is that in South Africa, property law is completely out of tune with human rights principles. In fact, far from property law being one of the foundations of human rights, it is one of the bastions of rightlessness. In feudal society, the serfs went with the land and owed duties to the landowner, but the landowner also had certain responsibilities towards the serfs. In South Africa, the feudal-type dependence exists without any corresponding obligations. It is the worst of all worlds.

There was a time, after the Anglo-Boer War, when it was Afrikaner farmers who became rightless on the land of their birth. Farms had been destroyed by the British Imperial Army, and fields belonging to so-called rebels were confiscated. Hundreds of thousands of poorer Afrikaner farmers became *bywoners*, dependent on the goodwill of the new legal owners of what they had once considered their ancestral farms. Their status was not all that much higher than that of so-called squatters today. The Afrikaner struggle over the land was part of the struggle over sovereignty, just as the campaign against poor-whiteism was an integral part of the battle for national and human rights for the Afrikaners. The big difference was that landlessness was not accompanied by votelessness. The vote enabled Afrikaners to restore their links to the land without having to question the principle of total obedience to legal title deeds rather than historical claims.

Looked at from a true human rights perspective, four groups would have claims on land presently reserved for white ownership.

The first are the black farmers of longstanding occupation who have never given up their insistence that they have rights to the land in question.

The second are those white owners who have rights recognized both in Afrikaner and African culture — by virtue of birthright, inheritance, occupation, investment, and work.

The third are those owners from the wider economy who have bought land and invested in its productive potential. Since their interests are essentially economic rather than proprietorial, their rights can be acknowledged in an economic rather than proprietorial form. The human rights aspect does not relate to the land as such but to their claim to have fair procedures to deal with their interests.

The fourth, and numerically possibly the largest, would be those from the wider economy whose ancestors were driven off the land by conquest, taxation, and hunger. Their claim lies not in rights to a particular plot, but in having access to land somewhere, either for habitation, or for farming, or both. Since the very system which expelled them from the land also denied them the possibility of acquiring sufficient capital to buy back the land, society as a whole would have a large responsibility for providing the means to enable them to recover rights to the land. For those who wished to resume farming, general criteria of justice and technical capacity would have to be established to guarantee that those most deserving got first place in the queue. Appropriate criteria would have to be applied to deal with access to housing and land for housing.

Nowhere is the indivisibility of human rights more evident than in the South African countryside. Violation of people's property rights has been accompanied by denial of general human rights; the restoration of the one cannot succeed without the recovery of the other. It thus becomes necessary to look at the wider legal dimension, and in particular at the introduction of the concept of constitutional rights and the Rule of Law for all who live on the land.

The legal dimension

In some ways South African law relating to land is too strong, in others it is too weak. It is far too strong and inflexible in its defence of the rights of title deed holders. It is far too weak in applying the Rule of Law and protecting basic human rights in the rural areas.

Law should aim to achieve five broad objectives for the countryside and the towns, irrespective of the government in power or prevailing economic policies, and irrespective of what principles and procedures may be used to settle land claims or govern land tenure and social organization. These objectives are:

☐ to protect the fundamental rights and liberties of all who live there;

☐ to extend the Rule of Law to prevent abuse of people's rights;

☐ to provide a minimum platform of social, economic, and cultural rights;

☐ to guarantee workers' rights; and

☐ to promote gender rights, combatting the oppression of women, and supporting the family.

These are all areas where the law today is at best silent, at worst an instrument of discrimination and oppression.

Extending constitutional rights

In South Africa we are not used to the idea of having constitutional rights. We are accustomed to the notion of power and counter-power, of pressures and permissions. While the courts and the press provide some checks and balances in high profile cases, they are rather marginal to the domination and abuse that infuse daily life. One fears at times that long after we have got rid of racism we will still have authoritarianism. Apartheid is dead. Long live authoritarian control. Nowhere is authoritarianism and arbitrary power over the lives of the people more evident than in the countryside. Nowhere is the sense of domination by some people over others more powerful than in the rural areas. Nowhere are the destinies of people so closely interwoven and yet so determinedly kept apart as on the farms. Children play together till a certain age, then one becomes the baas and the other a 'boy', the one a missus the other a 'girl'.

The extension of constitutional rights to the farming areas is the foundation of all other legal transformations there. It is only through the constitution that true equality of rights and dignity can be achieved in a multi-cultural, multi-faith society. The constitution does not, of itself, immediately eliminate the immense inequalities created by past racism, but it does establish a structure of equal political rights and equal protection under the law which enables the injustices of the past to be tackled. At the same time, the constitution speaks to all and for all. It is the biggest single agent for promoting the practices and habits of non-racism. Equality means that people who are different have the same rights. Non-racism is not one of South Africa's 'non-categories' whose identity lies in what it is not (like the non-Europeans, who, as one writer said, came from non-Europe). Non-racist means democratic, and implies taking people as they are and not attributing rights and duties to them on the basis of race. It is affirmative rather than negative, and acknowledging rather than dismissive of cultural variety.

The constitution is all-embracing precisely because it neither seeks to eliminate differences between people, nor pretends to eradicate social tensions and strife. On the contrary, the essence of the constitution is that it presupposes differences but states that they shall not be the basis for discrimination and inequality, and it acknowledges the inevitability (even value) of social struggle, but provides a framework within which it can occur peacefully and democratically.

Just as the first great constitutions were elaborated to overcome feudal absolutism, a constitution will be needed to prevail over absolutism on the land. It will not be a case of substituting one kind of racist absolutism with another, but of abolishing absolutism altogether, which will bring immense benefits to blacks on the land without stripping whites of their basic constitutional rights. Equality within diversity, not forced assimilation or forced separation, is the key.

People will at last begin to talk to each other as equals, not as master and servant, and within a framework of law rather than of arbitrariness. A common constitution is the basis of finding a common humanity and of establishing a shared interest in the land.

Application of the principles of the Rule of Law

This is part of the process of guaranteeing basic constitutional rights. In the first place, it means protection against arbitrariness and oppression. As has been said, whites do not only own land, they behave as though they own the people on the land. They exercise a private kind of control over the lives of others on the land, ignoring or respecting basic rights at their pleasure. The law sanctions evictions of people without legal process from their homes, acknowledges controls over the movement and visits of occupants that amount to localized banning orders, and permits every kind of disrespect in relation to the sensitivities and self-respect of persons on the land. Although the law does not expressly permit the use of violence against farmworkers, the atmosphere of domination and disrespect in the countryside makes any attempt to bring perpetrators of white-on-black violence to book a hazardous enterprise likely to provoke further aggression.

The application of the Rule of Law to the rural areas therefore presupposes new legal rights, a new kind of policing, a new magistracy, and new agencies for handling complaints.

Yet there is another way in which respect for the Rule of Law and legality can transform relationships in the countryside, and that is by upholding for the first time the binding and enduring nature of contracts freely entered into between different persons occupying and using the same piece of land. This is an aspect of legality that has great meaning to many black farmers, who to this day contest the validity of legislation purporting to override such contracts entered into by their parents and grandparents with local white farmers. The point they insist upon is that an agreement is an agreement until it is altered by mutual accord, and that accordingly its terms must be complied with even if they are inconvenient and even harsh. One of the great causes of the lack of legitimacy of the present-day system of land ownership is its failure to conform to contracts entered into by a previous generation of occupants of the land.

Acknowledging the legality and binding force of contracts has great relevance for correcting past injustices, since the existence or otherwise of local agreements would be one of the factors to be taken into account in determining who today would have specific historic and moral claims to a share in the land. Yet equally important, the drafting of contracts in conditions of equality, and according to objective criteria with due scope for subjective preferences, could permit an equitable sharing of the land that is non-disruptive of production and enjoys the respect of the people.

Legislation to guarantee a minimum platform of social, economic, and cultural rights

It should not be too difficult to work out a code of socio-economic rights for persons facing the specific problems of life in the rural areas. While voluntary activities will always be welcome, and religious, cultural, and social organizations will always have a role to play, they should act to supplement rather than replace guaranteed rights.

The code could deal with residential rights, treating farmworkers' kraals as homes rather than squatters' structures, and establishing appropriate protections similar to those given to protected tenants in the cities. In particular, this would protect the occupants against eviction except in very limited circumstances specified by law. It would also accord such homes all the rights of privacy and inviolability which normally attach to a person's domicile.

It could also provide mechanisms for the progressive opening of educational facilities, guaranteeing black children the same rights and responsibilities in relation to schooling that white children have, and making provision for adult education and literacy programmes. Until such time as free, compulsory, and universal education becomes the order of the day, there could be provisions catering for mixed contributions, that is, payment by parents, employers, and the state.

Similarly, the code could establish a framework of rights and institutions in relation to access to health services, also funded by mixed contributions pending the introduction of a national health system.

Another aspect the code could address would be access to learning and culture in the countryside. This could correspond with simultaneous steps to record, conserve, and develop local culture. In principle there is no reason why guaranteed access to television, radio, cinema, newspapers, and libraries should wipe out local culture. On the contrary, the stronger people are in their community culture, the more easily can they contribute towards and benefit from the culture of the nation as a whole, enriching and drawing on the cultural patrimony of the world.

Guarantees of workers' rights

Agricultural workers tend to be the poorest paid and most abused section of workers in the country. For agricultural workers in certain parts of the country, (such as employees on fruit or wine farms, whose aim is not so much access to land as better conditions of life for themselves and their families), the question of trade union rights, rights of collective bargaining, and the right to strike are fundamental. The law can provide mechanisms for registering agreements and settling disputes in the same way that it would for any economic activity, but give special attention to the specific problems related to agricultural work, such as the intimate relationship between employer and employees living on the same land, the seasonal character of work, and the fact that the workers live in homes tied to the land.

There is also a great need for a system of responsibilities and control in relation to safety, workers' clothing, and protection from the elements, as well as for guarantees of annual leave and sick leave.

It would not be the function of the law to spell out in detail the terms and conditions of every employment contract, but rather to guarantee certain minimum conditions, ensure that bargaining over contracts takes place in circumstances of fairness, and that all contracts of employment are reduced to writing, understood by both parties, and registered at some accessible place.

Recognition of women's rights and rights to a family and within a family

Patriarchy, racism, and feudal-type domination are all stronger in the rural areas than in the towns. Women thus face all the problems that their sisters have in the rest of the country, as well as special ones related to their particular situation.

The law has an important role to play in diminishing the isolation, silence, and abuse women in the countryside suffer. Their rights in relation to landholding and inheritance must be respected, and they should also be protected against eviction from the family home by selfish husbands or companions. Equal access of women to schooling, health facilities, and such social services as exist should be fundamental and guaranteed by law. Legislative attention should also be given to the problem of sexual abuse and harassment experienced by women who are especially vulnerable because of their isolation on the land.

Yet perhaps the most important right for women in the countryside is the right to participate in decision-making, whether in the home, in relation to contracts affecting their welfare, and at public meetings. Exercising this right could give women the confidence and authority to tackle their other problems.

The above are all areas where the law is woefully weak. There is one field, however, where the law is too robust. This is the concept of ownership, where rigid Roman Dutch principles of property rights stand in the way of flexible property arrangements based upon shared cultural values.

Interesting work has recently been done by lawyers at Afrikaans Universities with a view to distinguishing between ownership as *imperium* and ownership as *dominium. Imperium* is a feudal type of ownership which gives the title-holder control not only of the land but of the people on it. Roman Dutch law, as taken over and developed in the colonial type conditions of South Africa, emphasized ownership as *imperium*, so that by owning eighty-seven per cent of the land, whites accorded themselves the right not only to control the country's natural resources, but also the lives, activities, and movements of the millions of blacks living on their land. The suggestion has been made that the law be developed so as to separate the two aspects: *imperium* would then belong to the constitution and the organs created in terms of the constitution, while *dominium* would stay with the landowner in the form of proprietary rights and nothing more. The white farmer would continue to be the owner, but cease to be the *baas*.

This is a good beginning, even if expressed in Latin terms which few people understand, least of all those most directly affected by them. It opens the way to recognizing the fundamental human rights of occupants of land and ending the practice of regarding them as trespassers or squatters with no rights at all except the right to hope that the *baas* is in a good mood.

Yet it is only a beginning. The very concept of *dominium* or ownership is far too rigid to cater for the complex lattice-work of ownership rights that will be required when the land is one day shared in an equitable and pragmatic way. Re-legitimizing property law to take account of competing claims will require new concepts of ownership and new modalities of proof and registration.

The concept of ownership at present tends to be intolerant of shared or mixed control of the same property. The owner may be a company, or a partnership made up of many persons, or even a collective of persons possessing undivided shares, but ownership itself is single and undivided. Ownership has been described as the total cluster of rights the proprietor has in respect of the use and disposition of the property concerned. In the case of land, the owner is the person whose title is registered at the Deeds Office. In the case of movable property, other forms of proof are required. In the exercise of his or her rights of ownership, the proprietor of land may sell it, give it away, rent it out, use it as security for a loan and bequeath it to someone in a will. He or she may also grant a usufruct, that is, a right to benefit from the use of the land, which is a personal right lasting for the lifetime of the usufructuary. Another possibility is to burden the land with a servitude, such as a right of way, in favour of owners or occupants of neighbouring land. The categories are rather limited, arising out of the specific social and family needs of mediaeval Holland identified and fitted into Roman law categories by such great jurists as Grotius and Voet.

It is possible that the principles of lease and usufruct could be adapted, that irrevocable forms of trust could be developed to reflect new arrangements relating to concurrent interests in land. Principles of company law or partnership law could be introduced. Yet the law would become strained. The categories would not fit neatly, since their purpose was to deal with different situations. What seems to be indicated is an opening up of land law and an adaptation of its principles to ensure, firstly, that property rights are congruent with and supportive of human rights rather than in conflict with them, and secondly, that they correspond to the real situation on the ground and not to mythological or metaphysical notions connected to race (nor, for that matter, to the market or bureaucratic command).

The principles of Roman Dutch law in relation to ownership have in fact been regarded as immutable when applied against blacks, and as capable of infinite flexibility in response to the interests of whites.

Thus, the courts have had no difficulty in upholding the right of a white farmer to expel black occupants from his or her land, no matter that they and their families have farmed that land for generations, no matter that all kinds of arrangements intended to be binding were entered into between their grandparents and those of the present owner, no matter that they have nowhere else to go and no right or means to acquire land or shelter elsewhere, no matter that no public authority is under any duty to help them. At most, the more sensitive judges have insisted on a reasonable notice period ranging from some months to a year (what is reasonable, one wonders, after several lifetimes of occupation?). If one day the law were turned around and the ancient claims of whites were wiped out by statute, and the present owners referred to as squatters or unlawful occupiers, what indignation there would be at the violation of elementary property rights.

Far from Parliament attempting over the years to adapt the principles of ownership to the reality on the land, it has striven to compel reality to conform to the rigid principles of ownership. Thus the aim of statutes preventing blacks from owning or leasing land, or from entering into share-cropping or labour-tenant relationships, was precisely to combat the tenacious struggle of black farmers to retain guaranteed property rights, and to prevent any kind of sharing of interests in the land. Ownership, whiteness, and absolute control became synonymous, as did rightlessness, blackness, and subordination.

When it came to responding to the social, cultural, and commercial needs of whites in the cities, however, the law showed itself capable of infinite variety. A fundamental principle of Roman Dutch law has been that ownership of land and ownership of buildings could not be separated. From an ownership point of view, permanent structures were regarded as extensions of the land to which they were attached — one plot, one ownership. This made it impossible for persons to buy apartments, so the law was changed and sectional title permitted, that is, ownership rights could be registered in relation to parts of buildings separate from ownership of the land on which the building stood. For all practical and legal purposes, there was no longer any difference between ownership of a flat and ownership of a house. A fundamental principle of property law was violated to achieve this result, and found not to be so fundamental after all.

This was a question of dividing up the physical structure of property for the purposes of profitability and convenience. Then developers of holiday properties sought to find ways of dividing up time. On the principle that anyone can be rich for a week, they sought and succeeded in getting an alteration to the law so as to enable various persons to own the same property but at different times, the so-called time-sharing form of ownership.

At the more humble level, the law relating to landlord and tenant was moved away from absolute concepts of ownership and contract towards acknowledging principles of fairness and of accommodation rights. The contract between landlord and tenant, in terms of which the landlord could fix any rent he or she chose, was subordinated in specified cases to the principle of a fair rent determinable by the Rent Board in terms of objective economic criteria. The basic right of the owner to evict the tenant after the lapse of the lease period was subjected to the notion of protected tenancy, so that the tenant could stay on as long as he or she paid the rent.

What all these cases show is that property law is really what the white voters, and especially the rich ones, say it is. Now we are reaching the stage where we can start to envisage property law being what the voters, all the voters, rich and poor, black and white, say it should be. The search should accordingly be on to discover what the notions of property and property rights are that all the voters, or at least, the widest cross-section of them, would share in common. Sharing values is the first step towards sharing land and sharing the country. The question of involving the people in this inquiry thus becomes vital. The key to evolving a new land law is to discover what system of property rights would best correspond to the wishes and notions of the people. As the people change and their lives change, so their ideas change. Our task is to draw on the past, capture the present, and incorporate a capacity for development in the future.

With a view to enriching the debate, some ideas are offered in relation to the procedural dimension of the problem. If we can get the values, criteria, and procedures right, and if we can achieve agreement on them, the task of actually drafting the laws will not be so difficult.

Some tentative ideas will be advanced as to the kinds of procedures which could be adopted. This is an area where the remedies are as important as the rights. If we can talk to each other, if we follow democratic principles, if we search for solutions that are functional and manifestly fair in the circumstances, we will have taken major steps towards solving not only the land question but the people question. From being the major source of conflict and oppression, the land could over time become the foundation of establishing a shared belonging and the basis for a common patriotism.

The procedural dimension

Principles are interesting and procedures are boring, until your own interests are involved. Then principles seem less important and procedures matter. Clearly it would be foolish to attempt to lay down now what processes should be followed in re-distributing land, or rather, re-distributing rights to use of land. We do not know how political change will be accomplished, what role negotiations will play, what population shifts may occur, what the degree of physical confrontation may be.

Change brought about by an insurrectionary overthrow of apartheid could have very different implications from change through negotiation. Even if the principles of non-racial democracy were to apply, the procedures to deal with the effects of apartheid could be strongly influenced by the way apartheid is brought to an end. The emphasis in the following outline on the role of negotiated settlements on a farm-by-farm basis in terms of nationally agreed criteria, clearly presupposes a relatively peaceful transition in terms of which the spatial distribution of people will not be greatly affected. On the other hand, if white farmers prefer to die on the land or abandon it rather than share it, the proposals made below would have to be re-thought.

True sharing of the land, as in the case of true sharing of the country or of power, is not essentially a spatial or quantitative matter, an issue of quotas, but a question of values and interests.

When one speaks of the land, one speaks of the whole country, not just of farming areas. There has to be a comprehensive policy which takes account of all land, both rural and urban. The basic approach here is that the land belongs to all who live in it. There will be no racial hegemony over land, no concept that the land belongs to whites as such or to blacks as such. Many distinctions will have to be made in relation to different purposes to which the land will be put, and different procedures could be necessary in each case. One fundamental distinction is whether the rights under consideration are rights of necessity (living or survival rights) or rights of ownership (property rights).

Satisfying the rights of necessity involves finding land for homes, schools, hospitals, and recreational facilities. Appropriate procedures involving inquiry, public interest, expropriation, and compensation already exist and can be built upon and adapted bearing in mind the scale and urgency of the problem. The whole question of land rights in large and small towns requires separate and precise treatment, and will not be dealt with here.

The focus of this presentation is on ownership, more specifically on ownership of agricultural land.

The fundamental question will be to establish criteria based on shared values and capable of application according to specified procedures. To be effective, both the criteria and the procedures will have to be evolved through active consultation with farmers, both black and white. Clearly, all parties will look to the criteria that serve their own interests best. Nevertheless, it is not impossible to conceive of farmers agreeing to a kind of compact which corresponds to the realities of a country in transition, seeks to minimize unnecessary disruption, gives everybody something, and is consistent with widely accepted values in relation to property.

One can thus envisage a list of factors, each to be suitably weighted, based upon:

- birthright,
- occupation,
- productive use,
- inheritance, and
- title, both ancient and current.

The process of establishing these criteria will be of special importance, since it will in itself accustom people to share ideas and encourage the habit of looking for practical solutions based upon mutual advantage (even if only the limited advantage of preventing strife and mutual disadvantage). The idea would be to involve farmers throughout the country in a national debate with a view to thrashing out precise formulations. On the basis of this agreed-upon set of criteria, suitable legislation could be adopted. The same process would apply to legislation on procedures for using the criteria on a case-by-case basis.

The first phase

This would involve the symbolic and publicized return of recently expropriated land. The amounts of land involved here are relatively small, many of the areas concerned are still under government control, proof is easy, and the expelled communities can identify themselves without problems. The procedural and economic problems are relatively slight. At the same time, the emotional significance of such a restoration of rights would be enormous. Forced removals were the most vivid recent symbols of the subordination of property law to racist principles. They were amongst the most cruel representations of how the land question was tied up with the sovereignty question. They had no economic, social, or farming rationale other than to conform to the schemes of apartheid. They were strongly contested and highly publicized, with the whole nation involved in one way or another.

Facilitating the return of victims of forced removals to the countryside and creating conditions whereby they can live and farm in dignity would both acknowledge past injustice and indicate the beginnings of democratic solutions.

This rapid and unconditional restoration of rights would extend to all persons who had documentary proof of ownership or other real rights in land but whose interests were expropriated. Thus they might have had title deeds as individual owners of land, or held undivided shares, or enjoyed usufructuary rights, or had certificates of occupation, or been beneficiaries of trust deeds. Although the non-racial principle would apply, in effect all the persons benefiting during this phase would be black.

As the black farmers say: First give us back what we held until recently in terms of the white man's law itself and of which we have been robbed, then we can start talking about sharing the land as a whole with the whites.

The second phase

This would be the phase of stabilization and the creation of defensive rights in relation to land. It would also prepare the way for the next phase. It would involve legislation protecting occupants against eviction except on very limited grounds. It would create conditions of freedom of speech and organization on the land, guarantee basic trade union rights for agricultural workers, and create conditions for eliminating physical and human rights abuses.

Above all, it would be a period during which the criteria for evaluating claims to land could be hammered out, in which everyone on the land could begin to assert his or her dignity, and in which confidence could be established for moving on to the next phase.

The third phase

This would be the most difficult stage — making actual determinations. In areas with a tradition of sharing interests in land (even if on a totally unequal basis) consideration could be given to allowing the competing claimants to arrive at an agreement themselves, which, subject to proof of reasonableness and lack of oppression, could then be registered. For the main part, however, a land court or similar body would have to operate.

The land court could receive evidence from the claimants and then consider it in relation to the criteria as set out in legislation. The procedures would be such as to permit the presentation of all relevant material, with the bias of the court totally in favour of the legislatively enshrined values, and not towards one race or another. Compensation would clearly play an important role, particularly in relation to farms where there was no real possibility of sharing interests.

Thus, if a white farmer, born and bred on the farm and dedicated to its development, were to be awarded full title, black claimants would receive compensation, which could take the form of other land or of financial support for the acquisition of land. Similarly, where the application of the objective criteria resulted in an award in favour of a black farmer, the white farmer would receive appropriate compensation. The compensation would be calculated in terms of values and interests as indicated above, and not necessarily restricted to current criteria for assessing compensation in relation to land expropriated for public purposes. Clearly, many matters would have to be thought through before the governing legislation could be finalized. It would be crucial that the process be law-governed, orderly, and as fair and practical as possible in the circumstances.

If such a procedure could be properly worked out and applied, the result would be to satisfy blacks that their just claims were being recognized, while guaranteeing that the rights of white farmers with the strongest interests in and connection with the land, were legitimized and stabilized. The availability of resources for compensation would be of great importance in permitting flexibility of arrangements. The principle of the land being shared amongst those who work it could be realized in more ways than simply carving up a piece of land between various claimants, or giving it to those actually working on that piece of land, and no other (such as persons dispossessed of their rights at an earlier stage).

There is no simple solution to the land question, but the chances of success can be increased if a number of factors are addressed:

☐ The farmers themselves, both black and white, must be actively involved in the processes at every stage.

☐ While the criteria and procedures should be firmly based on well-established principles, their application in particular cases should be flexible and rooted in local reality, taking account of local patterns of land use and tenure.

While flagrant cases of forced removals should be attended to with relatively rapid procedures, other land claims should be handled in a particularly calm and sober way, starting with the areas and farms in relation to which relatively easy solutions can be envisaged and moving on step by step to the more difficult ones. The correcting of massive historic injustices cannot be done from one weekend to the next. It is in the interests of the farmers themselves, of the dispossessed and the possessors, and of the country as a whole that the process be an orderly and manifestly just one.

Postscript: the economic dimension

The argument:

The moral arguments for a human rights approach to the question of land ownership may be convincing, but any interference with existing patterns of ownership would be disastrous economically, especially if it meant dividing large modern farms and converting them into small plots for subsistence farming.

The response:

While it is true that most of the land is owned by whites, it is equally true that most of the actual farming is done by blacks. This central fact is frequently ignored when economic factors are placed in opposition to human rights considerations.

Those who have always talked and never listened must now begin to listen; those who have had to remain silent must now find their voices; creating conditions to enable farmworkers themselves to say what they want is fundamental, both to human rights and to production. The people living and working on the farms have the right to state what their demands are and how they would like to see them realized.

In the meantime, the experience of the incipient trade union movement on the land suggests that farmworkers in many parts of the country are not so much seeking ownership of the land as recognition of their rights as workers and as citizens.

If this is so, the effect of transforming relationships on the land in these parts of the country will rather undermine patterns of *baasskap* (domination) than patterns of ownership. The legal and psychological barriers to blacks acquiring rather than merely working on, say, Western Cape wine farms, would be removed, and attempts would be made to secure finance on favourable terms to ensure that no good farmer was prevented by race or other background factor from becoming an active part of the farming community.

Yet the main thrust of change in this sector of the rural economy, particularly in the earlier period, would be towards guaranteeing decent wages, household security, freedom from abuse, and access to education, health, and social amenities, rather than providing for new proprietary rights. Unequal relations of master and servant will be replaced by new relations based on freely negotiated contracts between employer and employee in a context of guaranteed constitutional and statutory rights. The kind and scale of the farming operations would remain largely untouched, but the rights of the workers would be greatly strengthened.

From another point of view, it could be argued that through judicious use of a land tax, land presently underutilized or badly utilized through lack of interest on the part of the owner, could be made available for farming. This would increase the amount of land available for re-distribution.

Yet the big question remains: Will the recognition of actual land claims by black farmers in respect of land presently monopolized by whites, lead to a parcelling-up of large productive farms currently producing for the internal and export market, and their replacement by small unproductive units dedicated to subsistence farming? Will social justice result in economic ruin?

It is difficult not to respond in emotional terms, whatever one's point of view. Yet it is precisely the questions that most arouse our passions that require the most sober appraisal and the greatest freedom from preconceptions, stereotypes, and mythology. Even if we all agree to base our analyses on objective reality, we have grave difficulty in achieving a common description of that reality. All the more reason for trying, bearing in mind that we should never re-shape or select facts simply to justify an already determined result, but rather aim to develop an analysis in which there is a logical and organic congruence between facts and proposals.

Some questions are so big that they break under their own weight, and a series of little questions burst through. In the case of what the impact on the rural economy would be of humanizing and legitimizing relations on the land, the not-so-little questions are: Is white farming economic? Is it white? Do black farmers want to divide up existing farms into small family holdings? Is black farming uneconomic, and if so, is that a result of inherent incapacity brought about by tradition, or does it flow from other factors associated with inequality and susceptible to rapid corrective intervention?

Is white farming economic?

If recent Department of Agriculture research is to be believed, only twenty to thirty per cent of white-owned farms are really productive in terms of yield potential. If the remainder give a reasonable income to the owners, it is because of direct or hidden subsidies. In the first place, there is the enormous and continuing debt that is sustained more for political than economic purposes. It has been stated that any government wishing to nationalize the land could do so without legislation and without compensation, by simply calling in the debt. One estimate is that in economic terms something like eighty per cent of the farms belong to the Land Bank, even if in legal terms they are registered in the name of white proprietors.

Then there is the question of subsidies. Prices are subsidized, particularly for export crops such as maize, which may not be the most economic in particular areas where it still pays to grow subsidized crops. It is one thing to subsidize agricultural production, it is another to subsidize life-style. It is one thing to encourage farmers to stay on and develop the land, it is another to facilitate absentee farming.

Productivity is indeed one of the factors which would be taken into account in determining the share that a farmer would get in relation to disputed land. It enters into the scheme of values that could be accepted by all as underlying the proposed new land law. Productivity is indicative of commitment to farming, application of sweat and intelligence to get the best out of the soil, and proof of having a relationship with the land. It is a factor acknowledged by black farmers, even in relation to white landowners with a record of bad treatment of workers. Where high productivity is coupled with birthright, the claims of the white farmer would be strong.

Similarly, there is land which is less fertile or less well-watered, and accordingly less productive, to which many white farmers have a genuine and active commitment. Such farmers would also have a strong claim, and any land tax would take into account the productive capacity of the land and not just its extent.

The farmers whose claims would be weaker would be those who absented themselves from the farm or who simply sat on the stoep drinking coffee while waiting to collect their subsidies, putting virtually no effort into improving the farm's capacity.

In all cases, what is contemplated is not so much an outright taking and re-division, as a computation of interests expressed in a legally protected form, acknowledging the new kind of shared access to and interest in the land, and allowing for flexible use of compensatory finance to facilitate equitable and pragmatically effective arrangements. Instead of being disruptive of the present productive reality, the new patterns of ownership would correspond more directly to it. There have already been far too many forced removals in South Africa, far too many expulsions from land, far too many refugees, to countenance a new trek of the dispossessed. What is envisaged is ending the system which makes such removals possible, stabilizing the rights of all who are on the land, and adjusting the terms of ownership and use to productive and cultural reality. The law of farms would thus move closer to the practice of farming, not further away from it.

There would be nationalization of land law, not nationalization of the land. An interesting suggestion has been made by an expert on mineral rights that rights in the sub-soil could be regarded as vesting in the state, as do rights to the sea-bed and rights to control the airspace. Surface boundaries make no sense in relation to underground mining. The supervision of the utilization of mineral resources could be equated with the responsibility of the state to support conservation of the land. It does not pre-empt the debate on ownership and control in relation to mining activity as such, but would facilitate acknowledgement of royalty rights for black farmers dispossessed of land on which mining later took place.

Is white farming white?

There is no way of testing the whiteness of farming activity as there is of washing with a certain powder or brushing one's teeth with a certain brand of paste. Yet the phrase 'white farms' gives the impression that they are farms owned and worked by whites, either using family labour or employing white farmhands. The actual situation is that there are invariably more blacks on the farms than whites, and there are many farms on which no whites live at all, leaving all farming to black managers and workers.

Blacks do the ploughing, the planting, the reaping, and the storing. They look after the animals, take on-the-spot decisions on how to respond to climatic changes and disasters. They are not just agricultural labourers, they are farmers, involved in the tradition and psychology (and, in some cases, in the practical bureaucratic aspects) of farming, but prevented by law from functioning as farmers on their own account.

Do black farmers want to carve up all the existing farms into small family holdings?

One of the greatest problems about the land is that decisions are continually being made without taking into account the values and the wishes of those most directly concerned. Full and extensive inquiries will have to be made before this question can be answered. We do not start with a clean slate, however. Farmworkers' leaders on the big agri-business fruit farms have indicated that what the agricultural workers on these enterprises want is improvement in their conditions of life rather than a breaking up of the farms.

Black farmers in the Western Transvaal have a tradition of working together on very large estates held in common according to principles of traditional law.

Black agriculturalists in other areas have become accustomed over generations to working on large farms, and would not necessarily prefer working on small units.

The important thing will be to ensure that the system of credits, prices, and rural extension facilitates the most economic forms of landholding, so that it is neither a question of big being beautiful, nor of small being beautiful, but of appropriateness.

Is black farming uneconomic?

One of the sad facts of South African history is that when African farmers proved their capacity and actually began to produce and sell better than the white farmers — as was the case when they marketed their produce in the newly opened diamond and gold fields — legislative and administrative means were deliberately introduced to destroy them as competitors. Recent studies by officials from the Development Bank and by researchers from Pretoria University have shown that in spite of all the legal and practical impediments placed in their way today there are black farmers in parts of the country who are achieving more sustained farming results than their white neighbours.

The approach suggested here is almost the exact opposite of searching the world for positive and negative models and then dragging them in to South Africa to prove or disprove a thesis. We prefer rather to look to other countries not for models but for options and experiences which may be useful to us in developing our own solutions. Among the relevant experiences in neighbouring countries are the successes achieved by cattle ranchers in Botswana and maize-growers in the formerly destitute so-called tribal trust lands in Zimbabwe, who, in conditions of political independence, made good use of state aid to produce vast surpluses of meat and grain for the market and for export.

These experiences need to be studied closely; their lessons may be debated, but what they refute is clear, namely the assertion that there is something in black social organization or culture that inhibits the production of large farming surpluses.

Low productivity in the bantustans would accordingly seem to be the product of destitution, oppression, and overcrowding rather than farming incapacity. While the pressure on the land is immeasurably greater than it was in Zimbabwe, precluding the possibility of a similar type of productivity breakthrough, there could be significant increases in production if the land were really de-racialized.

What is needed in South Africa is for the state to get off the backs of blacks and give them some of the supports until now reserved for whites. Then we will see who can farm and who cannot.

Conclusion to the postscript

No future constitution can ignore the land question. South Africa was not given either by providence or by conquest to any group. It belongs to all who live in it. The function of the constitution is to acknowledge this fundamental reality, not only by resonant preambular references, but by precise Bill of Rights formulations aimed at healing the tension between possession and dispossession. The Bill of Rights would thus have to deal with land ownership and with fundamental rights and freedoms of those living on the land.

It is not necessary or even advisable for the constitution to remain silent on the question of private property. Property rights have been so violated and the results so unfair, that a failure to attend to them would be a constitutional failure. It is not the function of the constitution to resolve every conflict, not even major ones. That is what Parliament and the courts are for. Yet the constitution must provide the framework within which disputes can be settled in a law-governed and fair way. This requires fair criteria and fair procedures against a background of common values.

Protecting rights to property therefore should be based on recognizing fundamental property values and not on preserving unjustly acquired property privileges.

A simplistic freezing of present patterns of ownership would militate against the general system of protected rights and freedoms contemplated for the rest of the Bill of Rights, and serve as a reminder that the sovereignty debate was far from over. There would be no agreed set of criteria for establishing rights to the land, simply assertion and counter-assertion. Whites would use their present position to build in opportunistic safeguards disguised as principles; blacks would use their future legislative strength to rectify the position. If no rapid solutions were found, peasants would simply seize property and ignore the constitution altogether.

If a constitution is anything, it is the expression of agreed values. What is urgently required now is the involvement of farmers and farm workers on a nation-wide scale in determining what these agreed values should be, or at least, what principles should guide the process of reaching consensus on them. The economic question is not the central one at this stage and should not, it is suggested, be allowed to dominate the constitutional debate. The crucial issue on the agenda right now is that of de-racializing and legitimizing land ownership and use, which, as has been explained, starts with scrapping the Land Act and the Group Areas Act, but goes well beyond that. The question of compensation itself becomes integrated into the process of objective criteria and fair procedures. The term 'just compensation' as used in Namibia requires precisely such a position.

Then, once farming becomes just farming, not black farming or white farming, and once ownership becomes just that, not control of people but control of land use, those who own and work the land will decide freely for themselves how they wish to be organized and what economic forms they choose to adopt.

The constitutional rights are accordingly, first, the right to legal acknowledgement of just claims to land according to objective criteria and fair procedures, and secondly, once the terms of ownership and use have been settled, the right of farmers to determine for themselves what to do with the land. The state neither insists upon nor excludes any specific economic form, but through its policies of supports and taxation encourages the production of food and gives special attention to assisting those deprived of opportunities in the past to overcome the disadvantages imposed upon them. At the same time, attention should be paid to involving people on the land, farmworkers and farmers alike, in developing the outlines of a charter of rural rights, which, in association with the constitution and the general law of the land, would ensure that the special problems relating to human rights, workers' rights, gender rights, and due process of law on the platteland, would be addressed.

Such an approach would be consistent with the broad sweep of the Freedom Charter, as well as with the Charter's specific provisions on land. It would carry the ANC Constitutional Guidelines a step forward, both in a substantive and a procedural way. What matters is not so much who authored the Charter or the Guidelines, as the principles they contain. One of the basic principles is that the people shall decide; this means all the people, not just this section or that. Another is that they will decide in conditions of freedom.

The solution to the land question depends more than anything else on the involvement of the people in the process.

10 Conservation and third generation rights: The right to beauty

South Africa has the possibility of becoming the second country in Africa to make conservation a constitutional principle. The first was Namibia.

Two major objections may be raised. The first is political. Is the question of the environment a truly national issue, of concern to the whole population and, as such, appropriate for constitutional treatment, or is it purely a sectional interest, even a diversion from the pressing constitutional problems raised by the transition from apartheid to democracy?

The second objection is technical. Many lawyers (in fact the great majority) would argue that however useful it might be to have legislation imposing duties to protect the environment, it is impossible to express environmental law in terms of positive rights. These lawyers would assert that human rights by their very nature belong to individuals, not to communities, and that it is quite unhelpful to speak of trees or animals having rights.

These arguments will be dealt with in turn.

Do you have to be white to be green?

> And well you might ask: why don't his poems speak of
> the leaves and volcanoes of his native land?
> come and see the blood in the streets
> come and see the blood in the streets.
>
> *Pablo Neruda*

It might appear irreverent to speak of the Maluti mountains and the rolling bushveld when blood is being spilt in our roadways; it would seem inappropriate to lament chimney-smoke pollution when the air is thick with teargas. People who have washing machines have no right to condemn others who dirty streams with their laundry; those who summon up energy with the click of a switch should hesitate before denouncing persons who denude forests in search of firewood. It is undeniably distasteful to spend huge sums on saving the white rhino when millions of black children are starving.

There are strong arguments against putting environmental rights on the already crowded agenda of struggle. At best, it appears tangential to the battle against apartheid, a fashionable idea imported with a six-month delay from Europe or North America. At worst, it is seen as yet another example of the imposition of the standards, tastes, and interests of the minority; a new form of establishing the superiority of the self-proclaimed civilized few (who allegedly care about nature and life's higher things) over the uncivilized masses (whose base habits, the corollary goes, have to be controlled).

Yet there is a deeper perspective that considers environmental concerns to be of national rather than sectional interest. Just as apartheid penetrates every aspect of South African life, so must the struggle against apartheid be all-pervasive; this struggle is first and foremost a battle for political rights, but it is also about the quality of life in a new South Africa. Apartheid not only degrades the inhabitants of our country, it degrades the earth, the air, and the streams. When we say *Mayibuye iAfrika*, come back Africa, we are calling for the return of legal title, but also for restoration of the land, the forest, and the atmosphere; the greening of our country is basic to its healing.

The emergent South African nation does not accept the idea that certain themes belong to one group and not to others. The theme of conservation must be prised free from restricted association with a certain section of the white community and be located within the desire of all South Africans for a cleaner and more decent life.

We share with all industrialized countries the problems of pollution, waste, exhaustion of resources, and destruction of ecological systems. These are the classic themes of environmental awareness. Even now, under apartheid, these issues have to be addressed. In a future democratic South Africa, they can be expected to receive even more attention. Health, the conservation of resources, and the protection of nature are of interest to all South Africans, particularly those preparing to be free citizens of a free country.

However, the recovery of South Africa from apartheid will require conscious advances on fronts that go well beyond the areas which in other countries have come to be associated with environmental law.

It is possible to present environmental questions as though they are unrelated to the country's history and geography, or to politically contentious issues such as apartheid. Indeed, they can even be presented as being of such transcendent proportions that getting rid of apartheid is a secondary and trivial pursuit. The answer, surely, is not simply to reverse the argument and say that the struggle for the vote and political rights is so transcendent that nothing else rates consideration. Instead, we must take responsibility for our country and all dimensions of its future, implanting the question of conservation firmly within our social and geographic reality, which includes the reality of struggle. Human rights in the broadest sense are indivisible. When we breathe the air of freedom, we do not want to choke on fumes.

Environment is a poverty issue.
No serious environmental programme for South Africa can ignore the question of poverty, nor can any system of environmental rights fail to deal with the right to accessible and clean water, to electricity or gas, to sewage and rubbish disposal.

Questions of environment are far more pressing for the poor than the rich. The rich take sanitation and domestic energy for granted. They are concerned about hidden menaces to their health — food additives, invisible fumes, and fibres. They worry about the invasion of alien bushes or creepers in natural parkland. They are anxious that rain might wash out their tennis or spoil their cricket. The poor have to dispose of their own waste. Water is frequently inaccessible, often unclean. They lack latrines. They have to collect firewood for heat. They are the main victims of drought and flood, of lightning and tornadoes. In a socially unequal society, relationship with nature becomes unequal.

Addressing poverty involves far more than increasing the disposable income of the poor. It necessitates the extension of utilities on a vast scale to reach all South African citizens. The right to light, heat, water, communication facilities, and waste disposal is fundamental to any environmental programme, not simply because it saves trees or streams, or reduces disease, but because people have a human right to enjoy such services.

Conservation is a workers' issue.
Environmental consciousness is of great importance to us all, but especially to workers. People are rightly concerned that pesticides can threaten the natural ecology and lead to contaminated foodstuffs, but they often forget that the most immediate victims are the workers who are expected to spray trees or handle toxic materials without protection. The guarantee of a safe and clean working environment

is a key question for any trade union, and integral to the broader question of a safe and clean South Africa. Workers are in a position not only to ensure that their own immediate environment is protected against hazards, but also to detect threats to the general environment at source, such as hospital syringes being dumped on waste tips where children play, or the passage of toxic effluent into rivers or the sea.

Conservation is a consumers' issue.
Not all South Africans are producers, but all are consumers, and share an interest in being protected against adulterated foodstuffs, wasteful packaging, and hazardous by-products. Radiation, carcinogenic dust and fibres, lead poisoning, and asthma-producing fumes, know no colour bar.

The environment is a cultural issue.
In the days before environmental consciousness developed, it used to be denigratory to say that someone 'came from the bush'. The English said it about Afrikaners, the Afrikaners about blacks. The fact is that the great majority of South Africans, both black and white, have until recently had close ties to the land.

African culture is layered with the insistence on viewing human existence as inextricable from its natural habitat. Trees, plants, animals, and humans co-exist in close interconnection. In terms of traditional religion and social custom, many animals were regarded as having a sacred or totemic character. Plants were freely used for healing purposes.

The animal kingdom was populated not by dumb beasts, whose only purpose was to await slaughter and consumption by humans, but by a living gallery of creatures whose expressive personalities were incorporated into the fables and legends of oral tradition. Although the hunt could be pleasurable, hunting was never purely for pleasure. Animals were killed to feed the hunters' families, not for the commercial value of hides, shells, and tusks. Wild life abounded, until the onslaught on the people and fauna of our continent began, and slaves and tusks were bundled into the holds of ships with equal unconcern.

The pastoral tradition, with its emphasis on human attachment to the soil and dependence on the climate, is deeply entrenched in Afrikaner consciousness. The sense of terrain is always there, in the vast journeys across the veld and over mountains, in the sanguinary battles, in the brave attacks of Boer commandos against British Imperial troops, in the bitter stories of drought and deprivation that followed the devastation of Boer farms. At school we English-speaking city-dwellers were forced to read stories in Afrikaans about a lion family or about troops of baboons; we far preferred tales of British pilots or French voyagers to the moon, but what I remember now that stuck in my consciousness and mingled with all my other deep South African memories during my years of exile, were the sensual images of landscape and the vivid stories of animal behaviour in Afrikaans books.

Today it is chic to consider yourself a walker on the beach or a person of the bush, living close to nature and respecting its rules. Not all that long ago the Strandlopers were decimated by imported disease and the so-called Bushmen were hunted with the same ferocity applied to the extermination of the natural fauna. Their sounds live on in the music of the country, and their imagination survives in the pictures of animals and humans painted on cave walls, an imagery that is deeply indigenous and speaks to all South Africans.

In her famous *Story of an African Farm*, Olive Schreiner gave a luminosity and texture to the landscape that located her characters in an unmistakably South African setting. Dreamer and philosopher of the veld, her deep roots in South African reality enabled her to take on and contribute to the great gender and social issues of all humanity. Sixty years earlier, the British settler Thomas Pringle, steeped himself in African social and farming life in the hills of the Eastern Cape; admirer rather than conqueror of Africa, he wrote with undisguised sympathy of the dignity of the amaXhosa and of the sensible and practical way in which they collected milk, extracted honey, and settled their problems.

Consciousness of nature and of our place in it runs deep in the many strands that make up the texture of South African culture. Until now, the struggle for the land has kept these visions apart as it has kept the people apart. Yet the possibility is there that a common sensitivity towards and love of our country in its physical and natural dimension will become one of the foundations of a shared love for the country in its social compass.

Do trees have rights?

Environmental law has emerged as a distinct subject, based largely on international treaties and domestic legislation. The question is whether ecological principles and the right to a safe and clean environment are appropriate for entrenchment in a constitution, and more particularly, in the new South African constitution-in-the-making.

South African constitutional tradition would be against it, but then it has been against many things found in constitutions elsewhere in the world, such as equality or a Bill of Rights. As has been argued above, environmental issues are national concerns, impinging on the quality of life of all sections of society. The question is whether these considerations can be translated into the language of rights, and if so, how they can be integrated into the constitution? To answer this question it is necessary briefly to examine the way the concept of human rights has evolved over the years, focusing especially on what has been called the three generations of rights (see Chapter 1).

One of the great gains of humanity over the ages has been to establish that every single person, irrespective of birth or background, has worth and dignity. The whole system of guaranteeing fundamental human rights has been built around the idea that each individual on earth has a set of basic and irreducible rights.

The so-called first generation of rights are the political, civil, and legal rights established by revolution against feudal and colonial absolutism in the eighteenth century. These 'blue rights' were posited on the notion of individual rights. All the classic civil rights struggles since then have been based on attempts by excluded groups to bring themselves within the ambit of these rights. Many people, in fact the great majority, were barred from constitutional protection because of race or gender. Great political battles were fought over access to the protection of the constitution — workers, former slaves, women, all had to struggle to prove that they too were the bearers of fundamental rights. The issue was not so much the nature of the rights, but the categories of persons who were to enjoy them.

By analogy, some people today extend these categories to include the rights of unborn children, animals, species, and more remotely, the rights of trees. They acknowledge that neither the foetus nor a whale nor a tree, nor even a whole rain forest exercising group rights, can bring actions to defend their rights — hence the concept of fiduciary or guardianship rights, enabling humans to bring actions on their behalf.

In my view, this approach is inconvenient and highly artificial. Instead of trying to fit all rights concepts into the classic mould of individual rights enforceable through the courts, a wider conception of rights should be developed to correspond with the new kinds of interests being protected. This has already happened in relation to the second generation of rights, the red rights.

The right to education, to health, to nutrition, to shelter, did not easily fit into the classic scheme of individually-based rights. What had previously been regarded as benevolent or charitable activities based upon moral or religious obligation, gradually became codified into law. Municipalities acquired a duty to provide clean water and collect rubbish, to build schools and hospitals. The concept of social, economic, and cultural rights began to emerge and is today firmly established.

Exactly how these rights materialized varied from country to country, yet they had one thing in common: unlike blue rights which projected the state as the potential enemy against which the rights of individuals had to be protected, red rights required public institutions to be the principal agency for their realization. While blue rights compelled the state to refrain from action, red rights demanded that the state deliver benefits and services.

Blue rights were not superseded, but complemented by red rights. The welfare state was not simply the extension of the Rule of Law to new areas, but something with qualitatively new characteristics. The welfare state established new legal institutions, new legal principles, new legal procedures, and a whole new structure of legality. Thus the right to safe working conditions ceased to be based on the right to sue a negligent employer in a court of law, and depended instead on a new system of factory inspectors and statutory duties supplemented by a completely different system of compensation in which commissioners and tribunals rather than law courts played a central role. Workers' rights did not cease to be rights because they were conceptualized and enforced in a different way; they were simply rights of a different kind.

Based on this kind of thinking, suggestions have been made for the development of a third cluster of rights, the so-called peoples' rights or rights of solidarity. They do not fit comfortably into either first or second generation rights schemes, and include such rights as the right to peace, the right to self-determination, the right to control over resources, the right to development, and the right to a clean environment; some might even include the right to information, the right to see the world, while others would also place gender-related rights and minority rights under this heading. Few would deny that these green rights are important, many would argue that they are not really rights at all.

There are strong arguments against considering them as rights. The most important one is that the assertion of vague, poorly-defined, and non-implementable rights undermines respect for genuine rights, and ends up diminishing rather than augmenting human freedom.

In reply to this proposition, it should be pointed out that rights have always evolved over time, both in terms of their substance and their modes of enforcement.

Some legal traditions emphasize that where there is a remedy there is a right, in other words rights come from the sovereign authority in the form of protections guaranteed through recognized institutional mechanisms. Others put the matter the other way around: first the right exists, then the remedy is found.

The reality is that there is constant interaction between rights and remedies. Rights express themselves through remedies, and yet are always straining against remedies; the idea grows faster than the institution. This is particularly relevant in relation to third generation rights. What start off as moral and political principles gradually evolve into legally enforceable rules. In the case of environmental rights, this process has been relatively rapid. Not long ago, environmental rights were the aspirations of a few dreamers. Soon, the concept entered the political debate, then came to be asserted as rights, and finally ended up as legally enforceable norms. International treaties and a vast range of domestic laws bear witness to this. The task we now face is to see whether certain fundamental themes can be discerned underlying this growing area of legal concern, and whether appropriate formulations can be found for inclusion in a new constitution.

For the reasons which have already been given, it is suggested that the themes of respect for the environment and of ecological systems are appropriate for constitutional recognition in South Africa. This could take many forms.

In the first place, the preamble could, amongst other things, establish the context of nature within which the South African people live and the common bonds flowing from inhabiting the same territory.

Secondly, there could be a clause in a section devoted to aims of the state (as in the constitutions of India and Namibia) which set out the objective of conserving the environment and furnishing all inhabitants of the country with safe and sufficient water and energy. Such a declaration would not be self-enforcing, but would serve as a point of reference for legislation and executive action, with special relevance in relation to interpretation of statutes, so that in the case of any doubt or ambiguity the interpretation favouring conservation should be adopted.

Thirdly, there could be procedural mechanisms requiring that due attention be paid to the environmental implications of any law or executive action. Thus a special parliamentary commission could be created to scrutinize and report on all legislation inasmuch as it affects the environment.

Fourthly, an environment ombudsman could be appointed to receive complaints from the public. Such person would have the power to report adversely on any governmental activity that represented a threat to the environment, leaving it to public opinion to force the government to correct its ways.

Fifthly, if the constitution were to deal with the question of land rights, it could contain special provisions relevant to the broad themes of conservation and the achievement of a sense of shared belonging.

In the first place, we have to establish the legal integrity of our common patrimony, the land we call South Africa. All restrictions on ownership and occupation of land based on race, gender, or ethnicity have to be ended. We need to become sons and daughters of the same soil, not of this or that homeland or bantustan or group area. Our country must achieve its true spatial dimension. Our enjoyment of and responsibility for it must be untrammeled by apartheid signs, group areas, or racial ghettos of any kind. This should be an express constitutional principle: South Africa belongs to all who live in it. The question of resources and control of renewable and non-renewable resources should also be given a constitutional foundation. Similarly, the principle of special protection for nature parks, green zones, mountains, and beaches should be expressly affirmed, as well as for special flora and fauna. Attention also will have to be paid to the whole question of what is meant by ownership (see Chapter 9), and in particular, to the responsibilities that accompany ownership. It would be disastrous if the constitution enshrined, directly or indirectly, concepts of absolute ownership entitling the titleholder to use or abuse the land as he or she saw fit. We need to establish the constitutional foundation for concurrent interests in land going beyond the presently-recognized categories, and also to provide a secure foundation of respect for the public interest in relation, for example, to development projects.

In addition to direct constitutional provisions dealing in general terms with questions of the environment, there could be an environmental code to establish binding norms on specific issues: pollution, waste, protection of species, soil erosion, protected zones, and so on. This code could be given a special status by coupling it to the constitution. Its terms would be more specific than the general provisions of the constitution, and accordingly, would require amendment more frequently than the constitution. The code could thus have a status half-way between the constitution, which could only be amended with considerable difficulty, and ordinary legislation, which could be amended by a simple majority in Parliament. It would be an accessible document of considerable educational value providing the framework for more detailed legislation and regulations in specific areas.

Finally, the constitution could acknowledge that our country is part of the African continent and that the question of ecology has both a regional and a continental dimension. The integration of a democratic and non-racial South Africa into the Organization of African Unity should open the way for the harmonization of South African experience with experience elsewhere in Africa and the creation of common institutions to deal with common environmental problems.

We return, then, to the question of whether trees or tree-snakes can be said to have rights. The above proposals suggest that plants and animals, like the air and the water, can be the object rather than the subject of rights. Conservation of the environment is a shared goal of all the inhabitants of South Africa. It is we who exercise these rights, in our own name and not on behalf of the flora and fauna. The interests we protect are our own rights to live in a clean and safe world in harmony with nature.

11 The consitutional position of whites in a democratic South Africa

What used to be called the black problem has now become the white problem. It is not easy to accept that even in relation to the demise of apartheid, whites and their anxieties dominate. Justice would require that the central issue be how to guarantee that the oppressed majority's rights are restored and the effects of centuries of colonial and racial domination removed. Negotiations should be exclusively about how to dismantle the structures of apartheid, establish democracy, and correct the injustices of the past. Yet what is being projected as a central issue is the constitutional future of whites.

In principle, this should be no problem at all: whites will enjoy full democratic rights like all other citizens. Whiteness will become a constitutionally irrelevant category. Those who are today classified white will cease to enjoy the special privileges that go with this attribution and become ordinary members of society. The fact that their whiteness disappears as a constitutional fact does not mean that they vanish as people. On the contrary, once the system of white supremacy is destroyed, their true interests as citizens, no better or worse than anyone else, can be protected, and this includes their interests both as individuals and as members of cultural, religious, and other groups.

What those who regard themselves as white and who are anxious about their future should therefore be demanding is guarantees that the constitution be democratic and that the fundamental rights and liberties of all be respected, without consideration of race, colour, gender, or creed. Yet what many are in fact asking is precisely that the constitution be non-democratic rather than non-racial and that every consideration be given to race, colour, and possibly to creed (the gender issue is simply too much for them).

Sometimes they claim to be speaking in defence of the rights of all minorities. On other occasions the issue is defined as preventing domination of one racial group over another. A further formulation is how to protect civilized standards, or, more fashionably these days, uphold first world standards.

What it comes down to is that whites at present control the whole apparatus of government and repression, they are given by law eighty-seven per cent of the surface area of the country, they completely dominate the economy, they have acquired the habits and culture of the master race, and they are reluctant to give up their privileges. At the more positive level, they are part and parcel of the history and culture of South Africa, they have skills and aptitudes which could be beneficial for the whole country, and in increasing numbers they are beginning to break away from racist ideas and practices.

They have the capacity to do enormous harm to the country, and also the possibility of transforming themselves as they take part in the process of transforming South Africa. Building a new nation in South Africa requires solving the white problem, that is, destroying the system of white supremacy and establishing the means whereby whites become ordinary citizens participating actively in the life of the society, neither more nor less privileged than anyone else.

There just cannot be co-existence between racial group rights and non-racial democracy. It would be like saying that just a little bit of slavery would be allowed, not too much, or that the former colonial power would exercise just a small amount of sovereignty over the newly independent state, not a lot. While the phased replacement of race rule by non-racial democracy can be contemplated, the constitutional co-existence of the two is philosophically, legally, and practically impossible.

A number of assumptions can be made about South African historical and cultural reality which are relevant to any constitutional proposals. The first is that the system of apartheid is unjust, hated by the majority of the population, and beginning to disintegrate under pressure. Second, South Africa is multi-lingual, multi-faith and pluri-political. Third, there are vast social and economic inequalities that have been established by apartheid laws and practices. Fourth, it is in the interests of all South Africans to prevent the collapse or serious impairment of productive capacity or public utilities. Fifth, there are certain universally accepted rights and freedoms which are as relevant to South Africa as to any other part of the world. Finally, the process of nation-building and overcoming past traumas will require constant and sensitive attention.

The argument that follows is that, difficult though the initial adjustment might be, a non-racial democracy in fact provides far more effective guarantees to whites (as to all South Africans) than does any system based on racial group rights.

Should a Bill of Rights be associated with federalism?

One of the main issues facing the National Convention that drafted the Union of South Africa Act was whether the Cape, Natal, Orange River, and Transvaal colonies should be brought together in a union or a federation. The initial assumption was that the new state would be a federation, but the delegates resoundingly opted for a union, insisting that there was no part of the sub-continent that was so separate in character, history, or economy as to justify any form of restricted sovereignty. The boundaries between the former colonies were accordingly dissolved, and provincial councils with delegated rather than exclusive powers set up. For nearly eighty years, to quote the official motto, Unity has been Strength. Only now that the prospect of universal suffrage is on the near horizon does Unity suddenly become Weakness.

As a matter of pure principle, there are arguments for and against union just as there are in relation to federation. A number of liberals in South Africa have over the years argued in favour of federation simply as a means of preventing over-centralization of power. The federal idea was coupled with the concept of a Bill of Rights. Had they left it at that and campaigned for universal franchise, their arguments might today have achieved considerable strength inside the broad anti-apartheid movement. The fact is that, with a few honourable exceptions, until quite recently they added a third check and balance that was manifestly racial in character, namely, that of a qualified franchise which effectively excluded the majority of blacks from the vote. The federal concept thus came to be associated with the paternalistic notion that blacks were almost, but not quite good enough to take part in governing the country.

The processes of state formation are manifold, yet as a general rule federations have come into being as a result of smaller states coming together to form a larger state rather than through larger states devolving power from the centre to regional entities. Thus the states that constitute the federation normally have an already existing distinctive historical and legal personality which serves as the repository for a residual or continuing sovereignty.

The problem that supporters of federation have in South Africa is that no such state entities exist, unless one dignifies the bantustans with either embryonic or residual statehood. The question of cultural rights and the future of traditional authorities can be dealt with in a way far more favourable to the people living in these areas than by imposing upon them an orphan statehood that would keep them poverty-stricken and cut off from the mainstream of South African development. In any event, most of them are far too fragmented to be seriously considered for any form of meaningful territorial autonomy.

From an economic point of view, South Africa has long been a common society. There are no autonomous or self-sufficient areas. The bantustans and the towns are closely if unequally tied by migrant labour and economic dependency. Over eighty per cent of the population, black and white, regard themselves as Christian. The ANC was formed in 1912 precisely to overcome tribal and regional divisions among the African people. Trade unions are national in scope. From the side of authority, the army, the police, the prison services are organized on a nation-wide basis; so are transport and telecommunications; there is one Stock Exchange for the country, one basic electricity grid, an integrated water supply system and a single time zone. Companies have one head office and even the sporting unions are national in character.

Drawing boundaries would accordingly be a highly artificial exercise. Far from corresponding to natural historical, cultural, and economic divisions, they would cut through highly integrated areas and populations.

In dealing with the whole question of federalism, it is useful to remember that different objectives might be wrapped up in the same concept. For some, federalism has merit because it prevents excessive concentration of power in any single authority and at the same time encourages respect for genuine regional differences. For others it is a way of depriving majority rule in South Africa of any meaning, by drawing boundaries around race and ethnicity. This would prevent the emergence of a national government, keep the black population

The constitutional position of whites in a democratic SA

divided, prevent any economic restructuring of the country, and free the economically prosperous areas of the country of any responsibility for helping develop the vast poverty-stricken areas. The issues are really ones of self-interest dressed up as principle.

As far as the former and more principled federationists are concerned, however, their anxieties about checks and balances could perhaps be met by other constitutional arrangements which did not have the effect of challenging the basic principle of equal citizenship in a united South Africa. The role of a Bill of Rights in the context of a separation of powers will be referred to later. At this stage it is worth mentioning that concern about the importance of maintaining grass-roots democracy and avoiding the emergence of an over-centralized and unduly bureaucratic state has come strongly from community-based sections of the anti-apartheid movement, giving rise to the possibility that strong forms of local democracy can be developed without dividing the country up into a myriad of political group areas.

Instead of posing the question: 'How can we weaken central government?', we may ask: 'How can we strengthen local government? How can we encourage direct community involvement, grass-roots empowerment, immediate accountability of those in authority? How can we promote the organic structures of civil society, and prevent the emergence of a remote, office-bound and potentially authoritarian state?' Such an approach would seek to harmonize rather than counterpoise strong local democracy with large national goals. It would help to detach the question of protection against over-centralization from the question of keeping people divided along racial and ethnic lines. It would also ensure that delegation or devolution of power was not a constitutional device for preserving enclaves of white privilege. Elements of the federal idea could be retained, but shorn of their entanglement with apartheid.

Protecting minority rights

International law and, to some extent, constitutional law, have long recognized that minorities have rights against majority populations. The right to dominate or to preserve special social and economic privileges has never been one of these rights, but the right not to be discriminated against has been, as well as the right to have one's language and customs respected. More recently, the right to specially favourable treatment to overcome the effects of past discrimination has received some measure of recognition.

Rigid, race-fixated structures would seem to be the worst means of securing these rights, even if only because they would induce massive distrust rather than trust, which is the foundation of the constitutional compact. Yet it should not be impossible to devise constitutional mechanisms consistent with the principles of non-racial democracy regarding the protection of minority rights.

Attention would have to be focused on what was meant by a minority. Here it is important to distinguish between defensive and affirmative, negative and positive aspects of being a minority. In a negative sense, any group singled out for attack or persecution could be regarded as a minority. If the basis of the attack or the means used are illegitimate, then all persons at risk have the right to claim legal protection. This is the principle of non-discrimination, which is the twin brother or sister of the principle of equal rights. In this sense, whites could be considered a minority group entitled to protection against any forms of harassment, abuse, or arbitrary action aimed at them because they are white.

The interests to be protected are identical to those of all other citizens, the negative interest of not being singled out on the basis of skin colour for unfavourable treatment. Whiteness has constitutional relevance in terms of its inappropriateness; it is relevant purely because it is irrelevant. It should not be used today as a justification for privilege and domination, nor should it be used tomorrow as the basis of humiliation and vengeance.

This is quite the opposite of saying that whiteness is a value in itself which merits constitutional regard. It is not the quality of being white that receives protection, but the quality of being human, of being a citizen. It amounts to saying that the constitution will not permit your whiteness from being held against you in any way; it is not declaring that your whiteness confers on you any special constitutional glow or immunity.

The situation is quite different with regard to groups identifying themselves by means of language, religious belief, or custom. While there is no such thing as a legal right to whiteness, there is such a thing as the right to belong to a group with a distinctive cultural or religious character. These rights are positive or affirmative. The extent to which they are recognized as express constitutional rights, with appropriate mechanisms for enforcement, varies from country to country.

The suggestion has been made that race classification for the purpose of voting be made purely voluntary, that is, that people adhere to the group with which they wish to identify, and that a special non-racial or South African category be created for those who do not wish to be classified. Apart from the absurdity of creating yet another group, this time the non-group group, or the non-racial race group, the so-called voluntary principle of group association does not deal with the question of how the groups are to be defined and what the criteria for individual admission are. If the groups define themselves purely on a subjective basis, then there would be no limit to their number. Any five or ten people could constitute themselves into a group and demand a separate voters roll. It could theoretically end up as one person, one vote, one roll.

If there is a limited choice of categories, people might object that none of them corresponded to the group with which they wished to identify. Alternatively, people might shop around from group to group, not in order to defend any historical or cultural interest, but to find the easiest way into Parliament. What would the criteria be?

If, as has been canvassed, one of the voluntary groups were to be that of Afrikaners, would that include only white Afrikaans-speakers, or also Afrikaans-speakers who today are classified as coloured and put on a separate voters' roll? Could the many Africans who use Afrikaans as their mother tongue qualify? Who would decide the principles, and who would determine in any particular case if an individual who sought voluntarily to associate with the group should be permitted to do so? Would existing members of a group be able to exclude new members? One can imagine the situation in which a candidate had a narrow majority and his or her opponent immediately sought to impugn the result by attempting to disqualify voters as not belonging to the appropriate category.

The idea of special votes and reserved seats is not totally unknown in constitutional law, but almost invariably it is conceived of in terms of a form of affirmative action to strengthen the representation of groups that historically have been grossly discriminated against.

What South Africans need above all is to acquire the habits and practices of living together, working together and voting together, and doing so as equals. A common voters' roll is the most fundamental indication of a shared citizenship and shared loyalty. It is the equivalent of independence for the former colonies. It is the mark of sovereignty, which for the first time will incorporate the people as a whole. Universal franchise on a common voters' roll will not in itself end apartheid, which is an intricate and all-pervasive system with institutional, economic, and psychological dimensions, but it will be both an historical acknowledgement of the fundamental equality of all South Africans, and the means whereby the inequalities and injustices of the past can be overcome in an orderly, law-governed, and democratic way.

There is no reason, however, why a democratic South African constitution should not contain explicit references to cultural rights, with appropriate machinery to guarantee their realization. The Freedom Charter has long recognized that cultural, linguistic, and religious pluralism are not only permissible but desirable — South Africa is a better country for being populated by people of many languages, traits, and creeds. Furthermore, the Charter implies that there shall be no single cultural formation or language or religious belief which shall be regarded as superior to any other. This would be particularly important in relation to language policy.

The use, say, of English as the language of international communication and of official business in the central government, would not mean that other languages would lose their equal status. On the contrary, the right to use one's mother tongue in Parliament or the courts, at the post office or in shops, should be guaranteed, as should the right to learn and develop one's language at school.

There is one area where the present realities of enforced group division might possibly be taken into account, and that is in facilitating the transition from white minority rule to non-racial democracy.

If one of the objectives is to encourage a common commitment of culturally and politically diverse groups to governmental institutions, then a sober analysis needs to be made of the fears and suspicions that keep them apart. From this point of view, what matters is not whether the fears that exist have their basis in objective reality, but whether they are genuinely held. Similarly, in this context the fact that those persons who are fearful are the very ones whose unconscionable behaviour in the past gave rise to the divisions and tensions underlying their apprehension, is only partially relevant. What is important is that the fears do exist and that they impede the movement towards non-racial democracy.

Temporary confidence-building measures based on acknowledgement of the current divide would not be inconsistent with the goals of democracy; on the contrary, provided their short-lived character was clearly understood, and the goal of non-racial democracy always kept firmly in mind, they could be seen as positive in the South African context. A caretaker administration based *de facto*, to some extent, on group or so called consociational forms of representation could be considered as one of many possible means for creating conditions for the introduction of full, non-racial democracy.

False protection of minority rights

There are non-racial as well as racial ways of entrenching white privilege under the guise of defending minority rights. The racial way is to construct the constitution around categories of race. The non-racial ways are two-fold; the one is constitutionally to freeze the economic and social status quo, the other to ensure constitutional protection for privatized apartheid. Both schemes fit in with the idea of Parliament being a place where blacks can talk as much as they like, but do very little, since the resources needed to bring about any major improvements would be constitutionally under white lock and key.

An apparently race-free clause in the constitution could protect vested interests from any governmental interference, or make the terms for intervention so onerous as to render change virtually impossible. As a result of the extraordinary coincidence previously referred to, the persons making this proposal happen to own eighty-seven per cent of the land and about ninety-five per cent of productive capital; they live in the best residential areas, send their children to the best-equipped schools, and have the best medical attention at their beck and call; they occupy all the top jobs in both the public and the private sectors and positions of command in the armed forces, police, and prison service.

With a view to encouraging an orderly and peaceful transition to democracy and minimizing the prospects of sabotage and disruption, provision could always be made for reducing the anxieties of those who fear that majority rule will drastically affect their standard of living and shatter their personal security. All sorts of arrangements could be made relating to pension rights (which seem to count well above the franchise and freedom of speech in the order of priorities), as well as to job security. A comprehensive programme could be worked out to prevent arbitrary seizure of assets, and deal with the question of compensation for property taken in the public interest. Appropriate mechanisms could be created to ensure that any such agreements were honoured.

None of this requires generalized constitutional treatment, however, except possibly in an explicitly transitional and short-lived way. If defenders of the grossly unequal and manifestly inequitable status quo wish to wrap their apartheid-derived privileges in constitutional provisions, one hopes at least they will have the grace not to do so in the guise of defending fundamental human rights.

The ultimate defence of white privilege would be to take it out of the domain of public law altogether and protect it as a private matter. This would be done by means of an apparently innocuous constitutional provision which simply acknowledged the inviolability of contracts and freedom of association or dissociation. If necessary the old Latin phrase *pacta sunt servanda* — agreements must be honoured — can be utilized.

Apartheid as a system of public law would be dead. The statutory division of the population on the grounds of race would be over. There would be no legalized discrimination, no official segregation of facilities, no racial group areas, no system of separate schooling, no apartheid in hospitals or swimming baths or golf courses. All that would exist would be a clause in the constitution permitting people to form private associations on a voluntary basis, and then another clause upholding freedom of contract. People could then get together and by virtue of pacts or restrictive covenants create racially exclusive residential areas, establish racially exclusive schools and hospitals and swimming pools and golf courses.

They would not have to start from scratch. They could merely club together and convert the whites-only group areas in which they presently live into whites-only residential suburbs. Any black person wishing to move into such an area and claiming that his or her constitutional rights to equal protection were being violated would be met by the defence that such protection only extended to the public and not to the private sphere. The constitutional right that would be recognized by the courts would be that of voluntary association and freedom of contract.

There would be some differences from the present position, the two most important being that the whites would have to finance any expenditure themselves and that individuals who did not wish to participate could refuse to do so. Yet the result would be to tie up massive resources in racially exclusive undertakings and to reproduce all the present patterns of a racially divided society, even if under a different legal guise.

The basic argument used to justify these schemes is that whites would be swamped by the black majority unless they received special constitutional protection.

It is contended that this argument is false, and that non-racial democracy offers a far more secure position for all South Africans, including whites, than do any of the special schemes.

White South Africans in a non-racial democracy

The virtues of non-racial democracy would seem to be self-evident in South Africa, and yet experience shows that they have to be spelt out. The basic scheme is a simple one. It represents the application in South Africa of universally held views and corresponds to the vision long projected in the Freedom Charter.

In essence, it presupposes a constitutional structure based on the following inter-related principles:

- [ ] equal rights for all South African citizens, irrespective of race, colour, gender, or creed;

- [ ] a government accountable at all levels to the people through periodic and free elections based on the principles of universal suffrage on a common voters' roll;

- [ ] political pluralism, a multi-party system, and freedom of speech and assembly;

- [ ] a mixed economy;

- [ ] protection of fundamental rights and freedoms through a justiciable Bill of Rights; and

- [ ] a separation of powers including an independent and non-racial judiciary entrusted with the task of upholding the Rule of Law and the principles of the constitution.

In the light of the pro-democracy upsurge in many parts of the world, such positions should be regarded as axiomatic and unassailable. Yet, against the background of what can only be described as racist assumptions, all manner of excuses are offered for departing from these principles in the case of South Africa.

For the purposes of analysis, it will be accepted that the prospect of majority rule, even if subject to a justiciable Bill of Rights, is alarming to the great majority of those who choose to classify themselves as whites in South Africa today. The argument will be that the best way to allay these fears is to ensure that democracy and its institutions are firmly planted in South Africa; the worst way is to undermine democracy from the start and subvert it with a complicated and unworkable set of institutions based on notions designed to keep racially defined groups locked in endless battle.

From a purely moral point of view, it is not easy to accept that the fears of the white minority in South Africa should merit special attention. It is they who made the bed in which they are now so unwilling to lie. If they are cut off from their fellow South Africans, it is because this was their choice. If they feel exposed because of their conspicuously high standard of living in the midst of much poverty, homelessness, and hunger, this is what they passed laws to maintain. If they are concerned at the tendency to solve political questions by force, they should recall that it was they who seized the country by forceful invasion, ruled it by force, and then outlawed peaceful protest and opposition.

Nevertheless, if we are to build a new nation on the ruins of apartheid, we have to address ourselves seriously to all the preoccupations of all the people, whatever their past roles. The abstract defence of democracy is easy; its concrete application is difficult, especially in a country where it has been much talked about and little practised.

When racists and democrats meet it is difficult for the racists not to be authoritarian and for the democrats not to be patronizing. Bearing that in mind, three areas will be selected for discussion on the basis that they are the most sensitive, controversial, and difficult. They are:

☐ loss of identity,

☐ collapse of the economy, and

☐ loss of freedom.

The question of identity: the right to be the same and the right to be different

Political rights and cultural rights

We are struggling in South Africa for the right to be the same. We are also fighting for the right to be different. No question has caused so much confusion as this one, perhaps because in the past the issues have been deliberately obscured.

The struggle for the right to be the same expresses itself as a battle for equal citizenship rights, as a struggle against being treated differently because one is black or brown or white or Christian or Muslim or Jewish or Hindu or female or male or Tswana-speaking or Afrikaans-speaking. We are all South Africans, human beings living in and owing loyalty to the same land. The country belongs equally to us all, and we belong equally to the country. There should be no differentiation whatsoever of citizenship or nationality between us. Nobody is worth more or less than anybody else because of his or her appearance or origin or language or gender or beliefs.

This is the principle of equal rights for every individual. In affirmative terms, it gives each South African the right to vote, to be educated, to travel, and to take part in the life of the nation. Expressed negatively, it is the right not to be discriminated against. No individual may be treated advantageously or disadvantageously because she or he belongs to a certain racial, linguistic, or religious group, or is of a certain gender. The protection applies not only to individuals but to groups; they shall neither be discriminated against nor shall they receive the benefits of discrimination against others.

The constitution must expressly and unequivocally guarantee the fundamental equality of all citizens, and establish appropriate mechanisms to make this guarantee a reality. The law must ensure that in all spheres of public life — education, health, work, entertainment, and access to facilities — no one is discriminated against because of colour, language, gender, or belief.

In South Africa today, physiognomy is destiny; your skin colour determines what your rights and duties are and how and where they shall be exercised. From a legal point of view, therefore, the struggle against apartheid is precisely a struggle against separateness and a struggle to be the same.

Sameness, however, should not be equated with identity. It is worth repeating that sameness relates to one area of life, identity to another. Sameness refers to one's status as citizen, voter, litigant, scholar, patient, or employee. In this capacity, one's appearance, origin, and gender are totally irrelevant. Identity relates to personality, culture, tastes, beliefs, and ways of seeing and doing things. Here we struggle for the right to be different.

The objective of non-racial democracy is not to create a society of identikit individuals, all looking the same, dressing in the same way, eating the same food, speaking the same language, voting in the same way, and doing the same dance steps to the same band (the so-called civilized person of earlier British assimilationist policy, who happened to be male, English-speaking, with a neat crease in his trousers, and a penchant for tomato sauce).

Equality, or the sameness of political rights, does not mean homogeneity or cultural blandness. As feminists and others have pointed out, to be equal in a hegemonic culture means to take on the culture of your oppressors. Non-racial democracy presupposes just the opposite. Political equality becomes the foundation for cultural diversity. Once the problem of basic political rights is solved, cultural questions can be treated on their merits. Liberated from the blockages and perversions imposed by their association with domination and subordination, the different cultural streams in South Africa can flow cleanly and energetically together, watering the land for the benefit of all.

The very concept of equality presupposes equal rights between those who are different. The aim is not to eliminate the different personal and cultural characteristics, not to get people to deny or be ashamed of (nor to over-glorify) who they are, but to ensure that these differences are no longer used for purposes of exploitation, oppression, insult, or abuse.

Language is a good example of an area where the principles of equality and diversity need to go together. No citizen should be entitled to more or subjected to less favourable treatment because of the language that he or she speaks; no language should be regarded as inferior or superior to any other language; there should be a policy of encouraging the development of South Africa's many languages.

Afrikaans writers and linguists have raised many questions about the future of the Afrikaans language in a non-racial, democratic South Africa. They are entitled to a clear answer from the constitution, bearing in mind that the ultimate guarantee is not that it is protected by the barrel of a gun, but that it is spoken by millions of South Africans, for whom it is the vehicle of their most intimate thoughts and feelings. Yet the question is not just how to secure the free exercise and development of Afrikaans, but how to guarantee full recognition of Zulu and Tsonga and Sipedi and all the South African languages disdainfully referred to as vernaculars, extending to them the status, dignity, and means for development that such recognition implies.

It will not be necessary for the constitution to attend directly to all the myriad problems associated with a democratic language policy. There will be questions relating to language use in Parliament, the courts, and the public service, in the police force and army, and at the level of local government. There will be the matter of medium of instruction at schools and universities, of the language of broadcasting, books, films, and newspapers, of place names and street signs. Special questions might arise in relation to languages spoken by smaller communities, or used for religious purposes, such as Gujerati, Portuguese, Greek, Arabic, and Hebrew.

The constitution would not necessarily have to respond to all these detailed questions, but it should frame broad operative principles, and indicate the mechanisms, including the courts, which will have the function of ensuring that these principles are adhered to.

Just as the constitution can guarantee language rights, so it can create secure space for free cultural expression in the broader sense of the term. Language is just one of the many factors that make persons regard themselves as members of a specific community. Religion might enter, as well as a variety of practices and traditions built up over the years and passed down from generation to generation. As has been said, folkways are often stronger than law ways. Frequently the transmission is unconscious, affecting such things as speech styles and body language. South Africa is the richer for having persons drawn from three continents, for being multilingual and multi-faith and pluri-political, for having pap and curry and roast beef, watermelon and grape.

The new South African constitution will accordingly favour diversity and an open society. It will recognize that the emerging South African nation comprises many different groupings with a multiplicity of languages and historical experiences. Cultural diversity and political pluralism are both desirable constitutional objectives. Each is important in itself, and each complements the other. What should be avoided at all costs, however, is the merging or conflation of the two. Basing political rights on cultural formation is to guarantee that the voting public will fragment itself into warring racial and ethnic blocs. It is also to ensure that true cultural expression is subordinated to shallow and opportunistic posturing of a chauvinistic kind.

The public domain and private rights

There is another dimension to the question of the right to be the same versus the right to be different, and that is in relation to where the public domain ends and the private sphere begins. In constitutional language, this means determining the point of intersection between the fundamental right to equal protection and the fundamental right to personal privacy.

We cannot imagine a constitution which seeks to prescribe whom people should marry or not marry, or whom they should have as their friends or dinner guests or companions. Nor should it permit any state official to dictate such matters. These are questions that belong exclusively to the individuals concerned, and the constitution will guarantee such rights of privacy. At the same time, a democratic constitution could not acknowledge a right to bar people from hotels or restaurants or taxis or sports facilities because of the personal prejudices of the managers. In the former case the right to privacy would take precedence, in the latter the right to equal protection would prevail.

Considerable experience exists in many countries on where to draw the line and what procedures should be followed in dealing with violations. It is clear that law by itself can do little to eliminate discrimination, but that in a climate of general social awareness and in a context of broad education in favour of equality, sensitively drawn legislation can play a significant role.

What would be disastrous in South Africa would be to convert the right to privacy into an instrument for permitting organized discrimination. The law should never be utilized as a mechanism for barring people from exercising their fundamental rights. It is one thing to say that the state shall never interfere with matters that are truly intimate and personal. It is another to say that the state should defend the right to exclude people from neighbourhoods or schools or jobs because they are blacks or whites or of Asian origin or Jews. This is an example of the situation where the right to be the same, that is not to be discriminated against, must override the right to be different.

The right to be different does not include the right to discriminate against others because they are different. Nor does it include the right to impose difference on others against their will. It is a right of personal expression that can be exercised by individuals and groups for their own well-being and satisfaction; it should never be used aggressively to curtail the rights of others.

La différence — the gender question

The question of the constitutional rights of women and men is dealt with in Chapter 5. Suffice to say at this point that the issue of the right to be the same and the right to be different would appear to be fundamental in any analysis. In terms of general political and civil rights, men and women have the right to be treated in the same gender-free way. The equal rights clause in a new South African constitution should be unambiguous in outlawing any discrimination or exclusion based on gender.

At the same time, many feminists argue that women are not simply men without penises, just as men are not simply women who cannot have babies. They want equality with men, but not necessarily according to the norms that men have created for society as a whole. Thus they do not want equality if that means they must be female men. True equality connotes a joint input into determining the generalized norms, as well as acknowledging the right of women and men to speak in their own voices.

There are others who put the emphasis on choice. Women and men should have the choice to decide whether to accept masculine or feminine roles, and they should not be penalized for either preference.

The related question of rights of sexual preference could also be tackled as one essentially of privacy and choice. On the one hand, there should be no discrimination against lesbian women and gay men because of their homosexuality (the right to be the same). On the other hand, their private behaviour is a matter for them alone and not for the state (the right to be different).

The question of property
The new ideologues

Once upon a time it used to be the dispossessed who were heavily ideological and the possessors who were flexible and pragmatic, at least so they claimed. Nowadays it is the property-owners who wish to submit society to pre-determined schemes, certain of their correctness as a matter of principle, and oblivious to evidence one way or the other.

People who traditionally favoured the establishment of a precisely-defined economic framework in the constitution, now argue for an open constitution which leaves the issue of economic policy to the wishes of future electorates and the good sense of future governments. Persons who formerly opposed any reference to economic matters in the constitution, now wish to load the constitution with economic clauses.

This group demands elaborate clauses in the constitution to protect private property, promote privatization, and entrench free market principles. In other words, after years of criticizing socialist countries for putting ideologically motivated programmes into their constitutions, and thereby removing the issues from public debate, they are now themselves planning to do just that, though from the opposite point of view.

What is at issue in South Africa today is not whether to have a market economy or a centrally planned one, capitalism or socialism. The basic problem is what to do about the fact that as a result of apartheid, whites today own eighty-seven per cent of the land and ninety-five per cent of the country's productive capital; that as a consequence of generations of legally segregated schools and hospitals, education and health services for whites are vastly superior to those for blacks; that in a country where tens of thousands of whites have private swimming pools, millions of blacks do not even have piped water.

Once the principle of a mixed economy is accepted, as it has been by all the major components of the broad democratic movement, the constitutional dimension of the issue falls away. What remains is the question of what to do about apartheid-induced inequality. Economic clauses apparently designed merely to guarantee the continuation of a system of free enterprise, actually have the effect of preserving a system of grossly unjust division of access to economic goods, that is, of guaranteeing much enterprise and little freedom.

At a constitutional level, then, the real issue is the competence of Parliament to deal with the totally skewed property relationships produced in South Africa by centuries of colonial dispossession and apartheid law.

The range of options is wide.

The constitution could sanctify the existing patterns of ownership and control, forbidding any public intervention at all.

It could, on the model of the European Convention of Human Rights, say nothing at all on the question, recognizing that it is a matter which permits of different views, and that ultimately the issue is one to be determined by the electorate.

On the other hand it could expressly permit the taking of property, but only in the public interest, and then subject to the payment of prompt and adequate compensation.

A variant of the last-mentioned scheme would be to allow intervention in the public interest, but to have a qualified form of compensation, in terms of which market valuation would not be the sole determinant. Affirmative action principles could enter the picture, so that, under broad equal protection principles, historical, social, and family factors could be taken into account, as well as the need to ensure continuity of productive use; there could be flexibility in terms of the modalities of payment, and a wide variety of transitional arrangements and forms of mixed interests could be permitted.

Finally, at the other end of the spectrum of possibilities, the constitution could expressly lay down a programme of economic reform, starting, say, with a declaration that the soil, sub-soil, and all the resources contained therein belonged to the state, which could then determine their use.

What seems to be clear is that simplistic global solutions are not useful, and that casting the issue in bald 'free market versus central planning' terms will not be helpful. In fact such polarization of the question at this stage merely facilitates evasion of the most pressing issues. There are indeed a number of highly relevant factors which cannot be slotted into the equation if put in this way.

There is the question, for example, of the way apartheid laws and practices have deformed not only the market but the whole area of entrepreneurial activity. Blacks have effectively been excluded as significant actors in the spheres of finance, production, and services. Backed directly and indirectly by the law, whites have exercised unconscionable degrees of monopoly control; trading has been manifestly unfair and racially-based restrictive practices have abounded. Far from barring steps to break this legal and *de facto* racist monopoly, the constitution should, in line with its general commitment to equal opportunity, facilitate them.

Then there is also the matter of the degree to which, even within the white community, economic control has been vested in fewer and fewer hands, with the result that today one study suggests that four-fifths of the shares quoted on the Johannesburg Stock Exchange belong to only four major conglomerates. The application of anti-trust legislation such as exists in the United States, where state agencies can compel the break-up of monopolies, could in fact have implications more dramatic than a drive towards nationalization. The new constitution would presumably permit and even facilitate the opening up of competition presently being blocked by this extraordinary degree of monopolization.

It is suggested, then, that the constitution should neither require nor foreclose specific economic policies. It is not necessary or even desirable for the constitution to be committed to any particular economic programme or philosophy. What the constitution should do, and this is the task of constitutions, is guarantee as much general fairness as possible, whatever economic policies are followed.

Fairness in South Africa would have three fundamental components:

☐ It would necessitate at least some degree of redistributive action to make up for past dispossession and discrimination, for example, special investment in housing, training, health, and education, as well as a policy to facilitate just access to the land.

☐ It would demand the opening up of the economy in the face of racial and other restrictive practices.

☐ It would require that any intervention be governed by law, be subject to the principles of public interest, and be controlled by manifestly just procedures, that is, that both the criteria and the procedures be just.

One further element can be added, namely that conditions for free and open debate on economic questions should exist. The context of respect for the fundamental rights and freedoms of citizens must be maintained, whatever economic policies are followed. Economic programmes should never be steamrollered through, but adopted after well-informed discussion.

If these principles are entrenched, the worst thing that could be done would be to block off any means of lawfully achieving redistribution. It is far more realistic and sensible, and in the interest of all South Africans, to aim at policies that acknowledge the need for structural adjustment away from apartheid, and then provide for manifestly fair procedures to accomplish this goal.

Guarantees could then comfortably be given that personal property, which has so much meaning in the day-to-day lives of people, would be immune to any form of expropriation other than that normally authorized by law; the principle of one person one vote could easily be supplemented by the principle of one person one home, one person one dog, one person one bank account, and so on, even if, as a constitutional norm, it did not attain one person one gold mine.

What about the workers?

It is not necessary to speculate about what workers' rights should be in a democratic South Africa — workers themselves are speaking, and a clear set of principles is beginning to emerge. South Africa has a long and complicated history of workers' struggles, the last decade having been particularly rich in experience. The demand has now gone up for the elaboration of a charter of workers' rights which could be adopted before or after a general Bill of Rights and which would consolidate the advances made by the workers especially in this recent period, and set out their perspectives for the future.

The possibility therefore exists of a hierarchy of legal provisions relating to workers' rights in a democratic South Africa. The foundation would be the constitution, which would guarantee the right to form trade unions, the right to collective bargaining, the right to strike, and possibly the right to a safe and dignified work environment.

In addition to reiterating these principles and giving them more precision, the workers' charter would itemize principles and procedures dealing expressly with a large range of issues such as equal opportunity, working conditions, safety, holiday rights, compensation for injury, pensions, gender-related matters, training and promotion, unemployment and dismissal. The charter could also contain clauses dealing with information that employers must provide workers, and the possibilities of workers being involved by enterprises in planning decisions.

As citizens, the workers would be able to campaign for socialism and support existing organizations dedicated to socialism, or form new ones, if that were their wish. The charter could re-state this right, or it could be left to the general clauses of the Bill of Rights, which would, of course, also permit workers or anyone else to campaign against socialism.

Depending on its form, the charter could be given an entrenched legal status, in terms of which it could not be amended as easily as ordinary legislation; it would serve as a point of reference for any specific legislation or executive action in relation to the matters within its purview. Any such legislation or actions would be invalid to the extent that it conflicted with the terms of the charter.

Finally, there could be specific statutes and regulations dealing with the concrete aspects of implementing the charter. These could all be collected together in the form of a code of labour law.

Affirmative action

In a sense we already have affirmative action in South Africa, but it is affirmative action in favour of whites. The state today spends about five times more on the education of each white child than each black child, and the disproportion in the sphere of health services is the same. The Land Bank advances billions of rand to white farmers in terms of loans that are not called in, while the amount available to black farmers is paltry. Figures have been produced to show that the inhabitants of Soweto are subsidizing municipal services for the luxurious white suburbs of Johannesburg.

Thus before we even arrive at affirmative action for the dispossessed, there is a lot of equalizing that can be done (in a sensible and orderly way, of course) simply by removing subsidies in favour of the privileged.

In essence, affirmative action in the normal understanding of the term, is a strategy which establishes a series of special efforts or interventions to overcome or reduce inequalities which have accumulated as a result of past discrimination. It acknowledges that the ordinary processes of law or of the market or of philanthropy or benevolence, are insufficient to break the cycle of discrimination, which replicates itself from generation to generation. Sometimes it is called positive discrimination, sometimes corrective or remedial action. The most widely employed phrase, however, is affirmative action.

The term affirmative action was invented in the United States in the 1960s to cover an important aspect of newly adopted Civil Rights legislation. Conceived as a means of materializing the principles of the equal protection clause introduced into the American constitution after the defeat of the slave-owning states in the Civil War, affirmative action programmes have had some measure of success in forcing employers to open up jobs and promotions to blacks, Spanish-speakers, and women. Their impact on education has been uneven, and in general one can say that affirmative action is highly controversial in the USA, with conservative forces generally being opposed to it.

Other countries which include affirmative action in their legislative programmes are India, where it operates to guarantee positions in public life to Untouchables and members of what the constitution calls native tribes, and Malaysia, where it functions as a mechanism for requiring the progressive transfer of control of economic enterprises from members of the minority Chinese population to the majority Malay community.

Clearly, affirmative action takes on different forms and has different meanings in different countries. South Africa could not simply take over experience from, say, the United States and expect it to work. The term is already being used today to cover pallid attempts to train and promote blacks within white-dominated enterprises. Useful though any advancement programmes may be, they fall far short of what affirmative action could mean in this country. In reality they are at present little more than the normal programmes of in-service training and promotion that any moderately forward-looking enterprise would undertake.

Considerable attention will have to be paid to the question of harmonizing affirmative action with non-racial democracy. Non-racism presupposes a colour-blind constitution; affirmative action requires a conscious look at the realities of the gaps between the life chances of blacks and whites. The right to be the same takes on an additional meaning — it is the right to have the same opportunities, and if these are blocked because of the heritage of past discrimination, then it includes the right to special intervention to remove the disadvantages.

In fact it is difficult to see how a truly non-racial society can be built in South Africa without at least one generation of accelerated progress being achieved under the principles of affirmative action. The promulgation of a non-racial constitution will clearly be vital, both at the symbolical level and in terms of guaranteeing equal political rights.

The constitutional position of whites in a democratic SA   171

Yet a non-racial society cannot be declared. It has to be built over the years, so that all vestiges of inequality on the basis of race and other criteria are removed.

Affirmative action, or some equivalent, will accordingly be required in the public service, in the security sector, in health, education, and housing; in relation to the land, and in respect of both the public and the private sectors of the economy.

As far as jobs are concerned, it is not a system of simply promoting blacks because they are black, but rather of making a special effort to improve the qualifications of blacks, so that standards of performance are maintained while the rich life experiences of all South Africans are brought into the work situations. In respect of land and entrepreneurial activity, it is not just a procedure for taking away from whites and giving to blacks, but one of working out comprehensive programmes of training, finance, and transitional arrangements, in which many legal forms are possible, ranging from joint ventures to purely private undertakings, to co-operatives, to village industries.

It would not be necessary to determine in advance in the constitution all the details of the scheme, or even whether there should be one comprehensive set of principles and institutions to cover the whole of South African life, or different sectoral arrangements. What could be laid down are certain principles which would govern the application of affirmative action as a modality for change wherever it was applied.

One such principle could cover the criteria justifying or requiring affirmative action in any particular area. Another could specify the importance of seeking solutions as close to the ground as possible. The principle of open hearings and of participation by all interested parties could also be established. The principle of the least onerous and disruptive remedy could be adopted; for example, if it were determined that within X number of years health facilities in a certain metropolitan area had to be extended so as to ensure equal service to all within that area, then those responsible for the improvements should opt for the least onerous method of achieving that goal.

The courts would not, then, analyse every programme in terms of its merits, but ensure that the proper criteria and procedures were followed in each case.

While there will inevitably be many points of overlap between the American and South African experience, there will be one central difference. In the United States, affirmative action inevitably has a paternalistic quality inasmuch as it relies upon sectors of the majority population agreeing to take action to open up possibilities for the minority. In South Africa, on the other hand, affirmative action will favour the majority of the population who can be expected to have strong positions in the legislature to back their claims.

One may fairly ask why it is necessary to have affirmative action at all if Parliament is there with the power through ordinary legislation to correct the injustices and inequalities created by apartheid. There would appear to be two good replies.

In the first place, it is necessary to clarify that the principles of equal rights and non-racism in the constitution should not be interpreted as mechanisms to block the elimination of the massive inequalities inherited from the past. Whites who have benefited so much from apartheid should not be able to come to court to complain that their constitutional rights are being violated because there are special programmes to deal with homelessness among blacks. (Possibly non-racial criteria could be found, so that affirmative action would favour groups defined in a non-racial way, for example, the homeless, the sick, the under-educated, the landless. In effect, but not absolutely and not as a principle, this would help blacks rather than whites, but would not be posited in that way. The problem would still exist at medium and higher levels of the economy or the public service, where it would be difficult to find race-free criteria for overcoming a situation that cannot be defined simply in social terms.) Affirmative action could be constitutionally recognized as a legitimate complement to the general principle of non-racism.

There is also the need to anticipate a tendency on the part of the judiciary to interpret the Bill of Rights as a conservative or blocking instrument designed to prevent any government interference with the status quo.

The concept of respecting vested interests is deeply rooted in South African judicial ideology. Indeed, it has been used on occasion to defend pockets of black land ownership against intrusion by the apartheid executive. In the absence of clear constitutional provisions, all future legislation and executive action designed to eliminate inequality in South Africa could be subjected to highly restrictive judicial scrutiny. The vast privileges vested in whites by forced removals, the successive Land Acts and the Group Areas Act, would be protected by the judges. Society would remain divided according to race, there would be a war of attrition between the legislature and the courts, and the constitution would fall into disrepute. Far better to draft the constitution in such a way as to make it clear that the peaceful and orderly elimination of inequality is one of its principal goals, and that all legislation should be interpreted with this in mind.

The constitutional position of whites in a democratic SA 173

The second reason for specifying criteria and procedures for affirmative action in the constitution (whatever its name and final form) is perhaps more fundamental. It is based upon a certain conception of South Africa and of the nature of any future constitutional dispensation.

It looks upon South Africa not as a country of majorities and minorities, each seeking selfish advantages against the other, but as a land of diverse people sharing a common humanity and embarking on the difficult road of establishing a common loyalty and patriotism.

Similarly, its vision of the constitution is not that of a document drawn up by victors over vanquished, nor that of a tawdry share-out of spoils between contenders for power on a fifty-fifty basis. It envisages the constitution as a solemn compact, a document based on trust and realism, which establishes in advance the ground-rules whereby all can live together in peace and dignity.

Such a constitution would enjoy the respect of all, since it would guarantee to the have-nots that there will be active moves to eliminate inequality, and to the haves that the process will be governed by law and operate according to manifestly fair and efficient procedures.

Conclusion — freedom for all

The one theme that unites all the above discussion is that the guarantees referred to are really not guarantees for whites at all, but guarantees for the whole population. This really is the guarantee of guarantees. What is being suggested is not a set of privileges for one section of the community to be defended by special constitutional mechanisms, and ultimately by force of arms or by outside intervention. Rather it is a constitutional arrangement created by South Africans for South Africans in a common determination to move away from the hatreds, divisions, and injustices of the past. A justiciable Bill of Rights becomes central to the defence of liberty for all.

It is in everybody's interest to feel free and at home throughout the length and breadth of the country. It benefits everybody to have the vote and the right of free speech and assembly and the possibility of throwing out a government that no longer commands respect. It is to everyone's advantage to be able to worship freely, speak one's language, and express oneself in the way one feels most comfortable. Everyone gains if the process of bringing about true equality is an orderly and peaceful one. The Rule of Law helps everybody.

This is really the guarantee of guarantees for whites, as for everyone else, namely that their deepest interests coincide with the deepest interests of their fellow citizens. What all South Africans should be trying to do is to strengthen the institutions of non-racial democracy, so that they become deeply implanted in the country and part of its general culture. Only in this way can the conviction grow in the whole population that the constitution is their shield, since it enshrines the principle at the heart of all democratic constitutions, namely that an injury to one is an injury to all.

12 Preparing ourselves for freedom: Culture and the ANC Constitutional Guidelines

We all know where South Africa is, but we do not yet know what it is. Ours is the privileged generation that will make that discovery, if the apertures in our eyes are wide enough. The problem is whether we have sufficient cultural imagination to grasp the rich texture of the free and united South Africa that we have done so much to bring about; can we say that we have begun to grasp the full dimensions of the new country that is struggling to give birth to itself, or are we still trapped in the multiple ghettos of the apartheid imagination? Are we ready for freedom, or do we prefer to be angry victims?

The first proposition I make, and I do so fully aware of the fact that we are totally against censorship and for free speech, is that we should ban ourselves from saying that culture is a weapon of struggle. I suggest a period of, say, five years.

Allow me, as someone who has for many years been arguing precisely that art should be seen as an instrument of struggle, to explain why suddenly this affirmation seems not only banal and devoid of real content, but actually wrong and potentially harmful. It is not a question of separating art and politics, which no one can do, but of avoiding a shallow and forced relationship between the two.

In the first place, repeated incantation of the phrase results in an impoverishment of our art. Instead of getting real criticism, we get solidarity criticism. Our artists are not pushed to improve the quality of their work, it is enough that it be politically correct. The more fists and spears and guns, the better. The range of themes is narrowed

---

Note: In July the ANC held an in-house seminar on culture. The organizers invited me to present a paper which would open up debate on the implications for culture of the ANC Constitutional Guidelines. The seminar took place in Lusaka, and the debate was lively.

down so much that all that is funny or curious or genuinely tragic in the world is extruded. Ambiguity and contradiction are completely shut out, and the only conflict permitted is that between the old and the new, as if there were only bad in the past and only good in the future. If one of us had the imagination of the Russian novelist Sholokhov, and wrote 'And Quiet Flows the Tugela', the central figure would not be a member of UDF or Cosatu, but would be aligned to Inkatha, resisting change, yet feeling oppression, thrown this way and that by conflicting emotions, and through his or her struggles and torments and moments of joy, the reader would be thrust into the whole drama of the struggle for a new South Africa. Instead, whether in poetry or painting or on the stage, we line up our good people on the one side and the bad ones on the other, occasionally permitting someone to pass from one column to the other, but never acknowledging that there is bad in the good, and, even more difficult, that there can be elements of good in the bad; you can tell who the good ones are, because in addition to being handsome of appearance, they can all recite sections of the Freedom Charter or passages of *Strategy and Tactics* at the drop of a beret.

In the case of a real instrument of struggle, there is no room for ambiguity: a gun is a gun is a gun, and if it were full of contradictions, it would fire in all sorts of directions and be useless for its purpose. But the power of art lies precisely in its capacity to expose contradictions and reveal hidden tensions — hence the danger of viewing it as if it were just another kind of missile-firing apparatus.

And what about love? We have published so many anthologies and journals and occasional poems and stories, and the number that deal with love do not make the fingers of a hand. Can it be that once we join the ANC we do not make love any more, that when the comrades go to bed they discuss the role of the white working class? Surely even those comrades whose tasks deny them the opportunity and direct possibilities of love, remember past love and dream of love to come. What are we fighting for, if not the right to express our humanity in all its forms, including our sense of fun and capacity for love and tenderness and our appreciation of the beauty of the world? There is nothing that the apartheid rulers would like more than to convince us that because apartheid is ugly, the world is ugly. ANC members are full of fun and romanticism and dreams, we enjoy and wonder at the beauties of nature and the marvels of human creation, yet if you look at most of our art and literature you would think we are living in the greyest and most sombre of all worlds, completely shut in by apartheid. It is as though our rulers stalk every page and haunt every picture; everything is obsessed by the oppressors and the trauma they have imposed, little is about us and the new consciousness we are developing.

Listen in contrast to the music of Hugh Masekela, of Abdullah Ibrahim, of Jonas Gwanga, of Miriam Makeba, and you are in a universe of wit and grace and vitality and intimacy, there is invention and modulation of mood, ecstasy and sadness; this is a cop-free world in which the emergent personality of our people manifests itself. Pick up a book of poems, or look at a woodcut or painting, and the solemnity is overwhelming. No one told Hugh or Abdullah to write their music in this or that way, to be progressive or committed, to introduce humour or gaiety, or a strong beat to denote optimism. Their music conveys genuine confidence because it springs from inside the personality and experience of each of them, from popular tradition and the sounds of contemporary life; we respond to it because it tells us something lovely and vivacious about ourselves, not because the lyrics are about how to win a strike or blow up a petrol dump. It bypasses, overwhelms, ignores apartheid, establishes its own space. So it could be with our writers and painters, if only they could shake off the gravity of their anguish and break free from the solemn formulas of commitment that people (like myself) have tried for so many years to impose upon them. Dumile, perhaps the greatest of our visual artists, was once asked why he did not draw scenes like one that was taking place in front of him: a crocodile of men being marched under arrest for not having their passes in order. At that moment a hearse drove slowly past and the men stood still and raised their hats. 'That's what I want to draw,' he said.

Yet damaging as a purely instrumental and non-dialectical view of culture is to artistic creation, far more serious is the way such a narrow view impoverishes the struggle itself. Culture is not something separate from the general struggle, an artefact that is brought in from time to time to mobilize the people or prove to the world that, after all, we are civilized. Culture is us, it is who we are, how we see ourselves and the vision we have of the world. In the course of participating in the culture of liberation, we constantly re-make ourselves. Organizations do not merely evince discipline and interaction between their members; our movement has developed a style of its own, a way of doing things and of expressing itself, a specific ANC personality. And what a rich mix it is ... African tradition, church tradition, Gandhian tradition, revolutionary socialist tradition, liberal tradition, all the languages and ways and styles of the many communities in our country; we have black consciousness, and elements of red consciousness (some would say pink consciousness these days), even green consciousness (long before the Greens existed, we had green in our flag, representing the land). Now, with the dispersal of our members throughout the world, we also bring in aspects of the cultures of all humanity. Our comrades speak Swahili and Arabic and Spanish and Portuguese and Russian and Swedish and French and German and Chinese, even Japanese, not because of Bantu Education, but through ANC Education. Our culture, ANC culture, is not a picturesque collection of separate ethnic and political cultures lined up side by side, or mixed in certain proportions; it has a real character and dynamic of its own. When we sing our anthem, a religious invocation, with our clenched fists upraised, it is not a question of fifty-fifty, but an expression of an evolving and integrative interaction, an affirmation that we sing when we struggle and we struggle when we sing. When we dance the toyi-toyi we tell the world and ourselves that we are South Africans on the road to freedom. This must be one of the greatest cultural achievements of the ANC, that it has made South Africans of the most diverse origins feel comfortable in its ranks. To say this is not to suggest that cultural tensions and dilemmas automatically cease once one joins the organization: on the contrary, we bring with us all our complexes and ways of seeing the world, our jealousies and preconceptions. What matters, however, is that we have created a context of struggle, of goals and comradeship within which these tensions can be dealt with. One can recall debates over such diverse questions as whether non-Africans should be allowed onto the National Executive Committee (NEC), whether corporal punishment should be applied at SOMAFCO (Solomon Mahlangu Freedom College), or whether married women should do high kicks on the stage. Indeed, the whole issue of women's liberation, for so long treated in an abstract way, is finally forcing itself on to the agenda of action and thought, a profound question of cultural transformation. The fact is that the cultural question is central to our identity as a movement: if culture were merely an instrument to be hauled onto the stage on ceremonial or fund-raising occasions, or to liven up a meeting, we would ourselves be empty of personality in the interval. Happily, this is not the case — culture is us, and we are people, not things waiting to be put into motion from time to time.

    This brings me to my second challenging proposition, namely, that the Constitutional Guidelines should not be applied to the sphere of culture. What?! ... A member of the Department of Legal and Constitutional Affairs saying that the Guidelines should not be applied to culture? Precisely. It should be the other way round. Culture must make its input to the Guidelines. The whole point of the comprehensive consultations that are taking place around the Guidelines is that the membership, the people at large, should engage in constructive and concrete debate about the foundations of government in a post-apartheid South Africa. The Guidelines are more than a work-in-progress document, they set out well-deliberated views of the NEC as enriched by an in-house seminar, but they are not presented as a final, cut-and-dried product, certainly not as a blueprint to be learnt off by heart and defended to the last misprint. Thus, the reasoning should not be: the Guidelines lay down the following for culture, therefore we must line up behind the Guidelines and become a conveyor belt for their implementation. On the contrary, what we need to do is analyse the Guidelines, see what implications they have for culture, and then say whether we agree and make whatever suggestions we have for their improvement. In part, we can say that the method is the message; the open debate the NEC wants on the Guidelines corresponds with the open society the Guidelines speak about. Apartheid has closed our society, stifled its voice, prevented people from speaking, and it is the historic mission of our organization to be the harbingers of freedom of conscience, debate, and opinion.

In my view, there are three aspects of the Guidelines that bear directly on the sphere of culture.

The first is the emphasis on building national unity and encouraging the development of a common patriotism, while fully recognizing the linguistic and cultural diversity of the country. Once the question of basic political rights is resolved in a democratic way, the cultural and linguistic rights of our diverse communities can be attended to on their merits. In other words, language, religion, and so-called ways of life cease to be confused with race, and sever their bondage to apartheid, becoming part of the positive cultural values of the society.

It is important to distinguish between unity and uniformity. We are strongly for national unity, for seeing our country as a whole, not just in its geographic extension but in its human dimension. We want full equal rights for every South African, without reference to race, language, ethnic origin, or creed. We believe in a single South Africa with a single set of governmental institutions, and we work towards a common loyalty and patriotism. Yet this is not to call for a homogenized South Africa made up of identikit citizens. South Africa is now said to be a bilingual country; we envisage it as a multi-lingual country. It will be multi-faith and multi-cultural as well. The objective is not to create a model culture into which everyone has to assimilate, but to acknowledge and take pride in the cultural variety of our people. In the past, attempts were made to force everyone into the mould of the English gentleman, projected as the epitome of civilization, so that it was even an honour to be oppressed by the English. Apartheid philosophy, on the other hand, denied any common humanity, and insisted that people be compartmentalized into groups forcibly kept apart. In rejecting apartheid, we do not envisage a return to a modified form of the British Imperialist notion, we do not plan to build a non-racial yuppiedom which people may enter only by shedding and suppressing the cultural heritage of their specific communities. We will have Zulu South Africans and Afrikaner South Africans and Indian South Africans and Jewish South Africans and Venda South Africans and Muslim South Africans (I do not refer to the question of terminology — people will determine that for themselves). Each cultural tributary contributes towards and increases the majesty of the river of South African-ness. While each of us has a particularly intimate relationship with one or other cultural matrix, this does not mean that we are locked into a series of cultural 'own affairs' ghettos. On the contrary, the grandchildren of white immigrants can join in the toyi-toyi — even if slightly out of step — or recite the poems of Wally Serote, just as the grandchildren of Dinizulu can read with pride the writings of Olive Schreiner. The dance, the cuisine, the poetry, the dress, the songs and riddles and folk-tales belong to each group, but also belong to all of us. I remember the pride I felt as a South African when some years ago I saw the production known as the Zulu Macbeth bring the house down in the World Theatre season in London; the intensely theatrical wedding and funeral dances of our people, performed by cooks and messengers and chauffeurs, conquering the critics and audiences in what was then possibly the most élite theatre in the world. This was Zulu culture, but it was also our culture, my culture.

Each culture has its strengths, but there is no culture that is worth more than any other. We cannot say that because there are more Xhosa speakers than Tsonga, their culture is better, or because those who hold power today are Afrikaans-speakers, Afrikaans is better or worse than any other language.

Every culture has its positive and negative aspects. Sometimes the same cultural past is used in diametrically opposite ways, as we can see with the manner in which the traditions of Shaka and Ceteswayo are used on the one hand to inspire people to fight selflessly for an all-embracing liberation of our country, and on the other to cultivate a sanguinary tribal chauvinism. Sometimes cultural practices that were appropriate to certain forms of social organization become a barrier to change when the society itself has become transformed; we can think of forms of family organization, for example, that corresponded to the social and economic modes of pre-conquest societies that are out of keeping with the demands of contemporary life. African society, like all societies, develops and has the right to transform itself. What has been lacking since colonial domination began, is the right of the people themselves to determine how they wish to live.

If we look at Afrikaans culture, the paradoxes are even stronger. At one level it was the popular Creole language of the Western Cape, referred to in a derogatory way as kitchen Dutch, spoken by slaves and indigenous peoples who taught it to their masters and mistresses. Later it was the language of resistance to British imperialism; the best MK story to appear in South Africa to date was written (in English) by a Boer — *On Commando*, by Denys Reitz — a beautiful account of his three years as a guerrilla involved in actions of armed propaganda against the British occupying army. Afrikaans literature evolved around suffering and patriotism. Many of the early books, written to find a space in nature to make up for lack of social space, have since become classics of world ecological literature. At another level, the language has been hijacked by proponents of racial domination to support systems of white supremacy, and as such, projected as the language of the *baas*. In principle, there is no reason at all why Afrikaans should not once more become the language of liberty, but this time liberty for all, not just liberty for a few coupled with the right to oppress the majority.

At this point I would like to make a statement that I am sure will jolt the reader: White is Beautiful. In case anyone feels that the bomb has affected my head, I will repeat the affirmation: White is Beautiful. Allow me to explain. I first heard this formulation from a Mozambican poet and former guerrilla, whose grandmother was African and grandfather Portuguese. Asked to explain Frelimo's view on the slogan 'Black is Beautiful', he replied: 'Black is Beautiful, Brown is Beautiful, White is Beautiful'. I think that affirmation is beautiful. One may add that when White started saying Black was ugly it made itself ugly. Shorn of its arrogance, the cultural input from the white communities can be rich and valuable. This is not to say the we need a WCM (White Consciousness Movement) in South Africa. In the context of colonial domination, white consciousness means oppression, whereas black consciousness means resistance to oppression. But it does establish the basis on which whites participate in the struggle to eradicate apartheid. Whites are not in the struggle to help blacks win their rights, they (we) are fighting for their own rights, the rights of free citizens in a free country, and to enjoy and take pride in the culture of the whole country. Whites are neither liberators of others, nor can their goal be to end up as a despised and despising protected minority. They seek to be ordinary citizens of an ordinary country, proud to be part of South Africa, proud to be part of Africa, proud to be part of the world. Only in certain monastic orders is self-flagellation the means to achieve liberation. For the rest of humankind, there is no successful struggle without a sense of pride and self-affirmation.

The second aspect of the Guidelines with major implications for culture is the proposal for a Bill of Rights that guarantees freedom of expression and what is sometimes referred to as political pluralism. South Africa today is characterized by states of emergency, restriction orders, censorship, and massive state-organized disinformation. Subject only to restrictions on racist propaganda and ethnic exclusiveness such as are found in the laws of most countries, the citizens of the South Africa envisaged by the Guidelines will be free to set up such organizations as they please, to vote for whom they please, and to say what they want.

This highlights a distinction that sometimes gets forgotten, namely the difference between leadership and control. Many of us are for ANC leadership; the organization's central position in South Africa has been hard won and the dream of the organization's founders is slowly being realized. Without doubt, the ANC will continue to be the principal architect of national unity after the foundations of apartheid have been destroyed and the foundations of democracy laid. Yet this does not mean that the ANC is the only voice in the anti-apartheid struggle, nor that it will be the only voice in post-apartheid South Africa.

We want to give leadership to the people, not exercise control over them. This has significant implications for our cultural work not just in the future, but now. We think we are the best (and we are!), that is why we are in the ANC. We work hard to persuade the people of our country that we are the best (and we are succeeding!). But this does not require us to force our views down the throats of others. On the contrary, we exercise true leadership by being non-hegemonic, by selflessly trying to create the widest unity of the oppressed and to encourage all forces for change, by showing people that we are fighting not to impose a view upon them, but to give them the right to choose the kind of society they want and the kind of government they want. We are not afraid of the ballot box, of open debate, of opposition. One day we will even have our Ian Smith equivalents protesting and grumbling about every change and looking back with nostalgia to the good old days of apartheid, but we will take them on at the hustings. In conditions of freedom, we have no doubt who will win, and if we should forfeit the trust of the people, then we deserve to lose.

All this has obvious implications for the way in which we conduct ourselves in the sphere of culture. We should lead by example, by the manifest correctness of our policies, and not rely on our prestige or numbers to push through our positions. Where political factors are relevant, we need to accept broad parameters rather than narrow ones: the criterion being pro- or anti-apartheid. In my opinion, we should be big enough to encompass the view that the anti-apartheid forces and individuals come in every shape and size, especially if they belong to the artistic community. This is not to give a special status to artists, but to recognize that they have certain special characteristics and traditions. Certainly, it ill behoves us to set ourselves up as the new censors of art and literature, or to impose our own internal states of emergency in areas where we are well organized. Rather, let us write better poems and make better films and compose better music, and let us get the voluntary adherence of the people to our banner ('It is not enough that our cause be pure and just; justice and purity must exist inside ourselves,' in the words of a war poem by Jorge Rebelo from Mozambique).

Finally, the Guidelines couple the guarantees of individual rights with the need to embark upon programmes of affirmative action. This too has clear implications for the sphere of culture. The South Africa in which individuals and groups can operate freely will be a South Africa in the process of transformation. A constitutional duty will be imposed upon the state, local authorities, and public and private institutions to take active steps to remove the massive inequalities created by centuries of colonial and racist domination. This gives concrete meaning to the statement that the doors of learning and culture shall be opened. We can envisage massive programmes of adult education and literacy, and extensive use of the media, to facilitate access by all to the cultural riches of our country and of the world. The challenge to our writers, musicians, painters, and dancers, to our dressmakers and potters and carpenters, to our broadcasters and journalists and publishers, to our teachers and sound specialists and film-makers, to all our cultural workers, is obvious.

13 The last word — freedom

We give the last word to freedom, yet we do not know what it is.

This is the central irony of the deep and passionate struggle in South Africa: it is for something that exists only in relation to what it seeks to eliminate.

We know what oppression is. We experience it, define it, we know its elements, take steps against it. All we can say about freedom is that it is the absence of oppression. We define freedom in terms of the measures we need to take to keep its enemy, tyranny, at bay.

Tyranny in South Africa means apartheid. That is the form that oppression takes. It is also the negative indicator of freedom; freedom is what apartheid is not.

When the call went up in the 1950s, 'Freedom in our lifetime', it signified the end of something very specific: colonial domination in Africa and apartheid tyranny in South Africa. The Freedom Charter adopted in 1955 was conceived of as the reverse of apartheid. A product of struggle rather than of contemplation, it sought in each of its articles to controvert the reality of the oppression people were undergoing. Its ten sections were based on the demands that a suffering people sent in, not on any ideal scheme created by legal philosophers of what a free South Africa should look like.

Any new constitution in South Africa must be first and foremost an anti-apartheid constitution. The great majority of people will measure their newly won freedom in terms of the extent to which they feel the arbitrary and cruel laws and practices of apartheid have been removed. Freedom is not some state of exaltation, a condition of instinctive anarchy and joy, it is not sudden and permanent happiness (in fact, some of the freest countries have the most melancholic and stressed people).

Freedom means being able to do what formerly was unjustly forbidden. If the majority of people can vote where they could not vote before, this will be freedom. If they can move as they wish, live where they want, feel at home everywhere in the country, this will be freedom. If they can speak openly and say what they believe, support the organizations they agree with, criticize those in authority, this will be freedom. If they can feel comfortable within themselves, have a declared pride in who they are, and a sense that they are recognized by the world they live in, then they will be free.

Freedom is indivisible and universal, but it also has its specific moments and particular modes. In South Africa the mode of freedom is anti-racist, and anti all the mechanisms and institutions that kept the system of racism and national oppression in place.

Yet if the anti-apartheid pre-condition is the foundation of freedom in South Africa, it is not on its own a guarantee of freedom.

The very thing that brings joy to the oppressed majority, namely, the end of the system they have always known, is exactly what induces apprehension in the oppressors. Those who traditionally have supported apartheid, and who today might concede, happily or reluctantly, that apartheid is wrong and doomed, are alarmed at what might happen to them when the structures they have lived by are destroyed.

The constitution has to be for all South Africans, former oppressors and oppressed alike. It expresses the sovereignty of the whole nation, not just a part, nor even just of the vast majority. If it is to be binding on all, it should speak on behalf of all and give its protection to all. In the past, rights for the one has meant tyranny for the other. Does that imply that the freedom of the oppressed can only be achieved by means of a new form of domination, this time of the majority over the minority, of black over white? Will freedom be guaranteed for all, or only for most South Africans? Or will the principle be followed that the constitution does not see majorities and minorities, as apartheid has always done, but only citizens, each as important as the next?

Secondly, the elimination of apartheid does not by itself guarantee freedom even for the formerly oppressed. History unfortunately records many examples of the freedom-fighters of one generation becoming the oppressors of the next. Sometimes the very qualities of determination and sense of historic endeavour, that give freedom-fighters the courage to raise the banner of liberty in the face of barbarous repression, transmute themselves into sources of authoritarianism and historical forced-marches later on. On other occasions, the habits of clandestinity and mistrust, of tight discipline and centralized control, without which the freedom-fighting nucleus would have been wiped out, continue with dire results in the new society.

More profoundly, the forms of organization and guiding principles that triumphed in insurrectionary moments, on long marches, in high mountains, that solved problems in liberated zones, might simply not be appropriate for whole peoples and whole countries in conditions of peace. These reflections have led some people to argue for inaction against apartheid because of their concern that removing one form of tyranny might lead to its replacement by another.

From a moral point of view, it seems most dubious to refrain from dealing with an actual and manifest evil because of anxiety that its elimination might lead to the appearance of another evil. Sufficient unto the day is the evil thereof — the best time for fighting for freedom is always now, and the best starting point is always here.

Usually those who claim to prefer the evil they know to the evil they don't know, come from a class that derives at least some benefit from the existing system. Oppression is for them something they hear about from others, something they dislike intellectually but do not suffer themselves. The possibility that they might be concrete victims in the future carries more weight than the fact that their fellow-citizens are being ill-treated today; sometimes, in a narcissistic way, they even cast themselves in the most tragic role of all, that of the helpless victim in the middle, powerless to affect events. For those suffering under oppression, on the other hand, the fact that there might be arbitrariness and abuses in the future counts for far less than the need to counteract the violence being done to them today.

In any event, whatever the stand-point, the question of guarantees of freedom for all is an important one that needs to be confronted now. It has a bearing both on the character of the constitution and the process whereby the new constitution is forged.

There can, of course, never be absolute guarantees in history. What we do know for sure is that attempting to defend minority privileges by force of arms, whether through the present system, or by means of a constitution based on group rights, can only result in continuing strife and violation of human rights. The only system that has a chance is one based on non-racial democracy. What we need to do is strengthen the prospects of non-racial democracy being brought about as swiftly, securely, and painlessly as possible.

A democratic constitution is one integrated entity. It does not have 'own affairs' sections — one set of guarantees for blacks, another for whites. A constitution is a document with an intellectual reach into the future. It is our generation that drafts it in the light of our historical experience and the thought of our age, but we consciously attempt to produce something that will last. If we wish to break down the habits of thinking in racial categories and to encourage the principles of non-racial democracy, we must produce a constitution that contemplates the rights of all the citizens of our country, not just of a section, however large and however abused in the past it may have been.

To be effective, the constitution must be rooted in South African history and tradition. It must draw on the traditions of freedom in all communities, not just those who at this historical juncture are in the forefront of the freedom struggle.

There is, in fact, no section of the population, whatever its position today, that has not at some time in its history fought for freedom. Many of the foreparents of the whites who live in the country today were refugees from persecution, the Huguenots who fled from massacre because of their religion in France, the Jews who escaped from pogroms and then from Nazi terror. Thousands of English-speaking whites presently occupying important positions in the professions and public life, volunteered for military service against Nazism and fascism in Europe and later marched in the Torch Commandos against the extension of racist rule in South Africa.

South Africa has had an unusually large number of bishops who have been willing to go against the tide, usually stronger in their own churches than outside, as well as of writers and journalists and lawyers and academics and medical people (even at least one freedom-fighting dentist and two road engineers).

There is not an Afrikaans-speaking white family that was not touched by the struggles over the right to speak Afrikaans and have an Afrikaner identity; Boer heroism against the might of the British Empire became legendary throughout the world, and is part of South African patrimony, just as the concentration camps in which thousands of civilians died are part of our shame.

Workers from all over the world, driven by hunger and unemployment, came to work on the mines in South Africa, where they died in huge numbers of lung disease; hundreds fell at the barricades, gun in hand, as they fought against reduced wages; the tradition of patriots singing freedom songs as they face execution was started by four trade unionists who sang the Red Flag as they mounted the gallows.

Many South African women joined the suffragette movement and challenged the physical, legal, and psychological power of male rule.

Apartheid has distorted this history, subordinating each and every action to its racist context, suppressing all that was noble and highlighting all that was ugly. The ideals of democracy and freedom are presented as white ideals, the assumption being that blacks are only interested in a full stomach, not in questions of freedom. Daily life refutes this notion.

It is the anti-apartheid struggle, not the white presence, that has kept democracy alive in South Africa. The term 'anti-apartheid' has come to mean 'pro-democracy' in South Africa. The principles of non-racial democracy have for decades become part and parcel of the anti-apartheid movement, and through it have emerged as strong themes in South African life. This is not only borne out by the number of organizations that support a document such as the Freedom Charter, but by the growth of a powerful, alternative, democratic culture in the country. The culture of democracy is strong precisely because people have had to struggle for it.

In the last resort, the strongest guarantee of freedom in South Africa lies in the hearts of the oppressed. It is they more than anyone who know what it is like to have their homes bulldozed into the ground, to be moved from pillar to post, to be stopped in the streets, or raided at night, to be humiliated because of who their parents are or on account of the language they speak. Inviolability of the home, freedom of movement, the rights of the personality, free speech — they fight for these every day. If the constitution is suffused with the longing of ordinary people for simple justice and peace, then freedom in South Africa is ensured.

It could have been otherwise. There could have been a movement which accepted the racist premises of apartheid, but simply reversed the roles. Instead, the anti-apartheid movement based itself on establishing a better and more moral system than the one it dedicated itself to overthrow. The ideals of democracy were nourished in the hard soil of Robben Island, in the underground, in exile. They were taken up by the churches and other religious bodies, they were integrated into the life of the trade union movement. Journalists, lawyers, teachers, doctors, and nurses challenged apartheid with democratic ideas.

Thousands of community organizations were established throughout the country with a view to creating democracy at the grass-roots level. A great deal of experience was gained during this period, a great part of it positive, some of it negative. It has all been discussed, theorized about, argued over. People are more aware than before of the immense possibilities and also of the dangers of exercising power at the local level. Mistakes have been made and cruel things done in the course of the struggle, but there has never been any acceptance of the idea that the viciousness of apartheid and the nobility of the democratic idea permit the use of vicious means in the fight against oppression.

Constitutions can have many meanings. In the first place, they establish the structures of government, and lay down how political power is to be exercised. Yet a constitution does much more than outline the political and legal organization of the state. It serves as a symbol for the whole of society, as a point of reference for the nation. People like to feel that they have constitutional rights even if they do not exercise them. The existence of a constitution is an indication that society is ruled by steady and known principles of law and not by the arbitrary whims of persons. Like the flag, the anthem, and the emblem, the constitution stands above everybody and everything and symbolizes a shared patriotism binding on all.

The constitution can also serve as an educator. Its language is invoked in all sorts of situations, it is studied in school, it becomes integrated with the general culture of the society. The language of freedom in the constitution becomes part of the discourse of the people.

In South African conditions the constitution will in addition be a compact or agreement, solemnly entered into by democratically chosen representatives of all the people, emerging out of strife, with the sense of and commitment to the creation of a set or rules in terms of which all can live together with pride and in peace.

Above all, the constitution is a vehicle for expressing fundamental notions of freedom, at the conceptual, symbolic, and practical levels. In South Africa this aspect has special importance. An effective Bill of Rights can become a major instrument of nation-building. It can secure for the mass of the people a sense that life has really changed, that there will be no return to the oppressive ways of apartheid society, while at the same time giving those who presently exercise power the conviction that their basic rights will be guaranteed in the future without recourse to group rights schemes.

It will be one constitution with one generalized set of provisions guaranteeing basic rights and freedoms to all. Some might look with special interest at the sections dealing with freedom from fear. Others might focus on the question of freedom from want. Many would be concerned with the third great freedom, namely freedom from insult.

Each set of provisions will be important in itself. The classic civil, political, and legal rights — the so-called first generation of human rights — need to be autonomously defended through the classical mechanisms of elections, free speech, and judicial review. The second generation of rights — social, economic, and cultural rights — are no less important. They too will be upheld by appropriate mechanisms, in which Parliament will play a key role.

The right to be free and the right not to be hungry are both fundamental human rights to be defended and fought for as vigorously as possible. One cannot permit the existence of the one to negate or diminish the importance of the other. The fact that there needs to be a great national effort to combat hunger and homelessness is no reason for cutting back on freedom of speech or the right of access to the courts. Similarly, the fact that citizens can consult their lawyers and get a court order in their favour in no way mitigates the need to provide a legal framework to combat hunger. The constitution is not unfriendly to private philanthropy, but does not see it as a substitute for the progressive materialization of rights.

Similarly, the third generation of rights, namely the rights to peace, development, and respect for the environment, will also be integrated into the constitution. It is only in recent years that these have begun to crystallize as legal rights, and much still needs to be done to provide appropriate formulations and remedies. We cannot expect an elephant to apply for habeas corpus, but a generalized principle of interpreting all laws in a way that favours conservation, and the imposition of ecological duties on local authorities, as well as the creation of a citizen's remedy as pioneered by the Indian Supreme Court, could meet the situation.

For many years, supporters of majority rule looked with suspicion on the idea of a Bill of Rights and the Rule of Law. On the other hand, proponents of entrenching fundamental rights and freedoms balked at the notion of one person, one vote. Now, these two currents — that tended to flow in different directions — have converged. The resolution of questions relating to political rights and fundamental liberties makes it possible to give guarantees on the rights of cultural diversity. In combination, these concepts can ensure manifestly fair procedures for regulating the process of eliminating the inequalities created by apartheid.

Spelt out in terms of constitutional principles, one can envisage the following clusters of entrenched guarantees:

- [ ] The constitution will be designed in such a way as to ensure full and equal participation in political and civil life for all South Africans, irrespective of race, colour, gender, or creed.

- [ ] Discrimination on the basis of race etc. will be outlawed, and machinery created to prevent insult, abuse, or ill-treatment on such grounds.

- [ ] There will be a multi-party system with freedom of speech and assembly and periodical elections to choose Parliament and the government.

- [ ] There will be a Bill of Rights guaranteeing fundamental human rights and liberties to all citizens. This Bill of Rights will be entrenched in the constitution and will be justiciable, that is, persons alleging infringements of their rights will be able to seek a remedy by recourse to the courts. Provision should be made to ensure equal access to the courts independently of financial means.

- [ ] The application of the doctrine of the separation of powers will establish a system of checks and balances between Parliament and the executive, and guarantee that the judiciary is independent in fulfilling its functions of upholding the Rule of Law and defending the principles of the constitution.

- [ ] Steps will be taken to ensure that there is vigorous government at the local and regional levels in line with the principles of permanent accountability and active community participation.

- [ ] Within the context of a single citizenship and a common patriotism and loyalty, the diversity of the South African population will receive constitutional recognition through provisions guaranteeing the free expression of religious, cultural, and linguistic rights.

- [ ] The opening up of equal opportunities for all and the process of redistribution of wealth in the country will be conducted according to constitutionally defined principles covering public interest, affirmative action, and fair procedures, with the courts having the power of judicial review in relation to the defence of these principles.

These are not provisions for black South Africans or for white South Africans, but for all South Africans. The last word goes to freedom.

Appendix 1
Two underlying questions

Is South Africa an independent state?

Legal truth, like all truth, arises out of the clash of opposites. At the heart of the debate on the legal characterization of the apartheid state are two seemingly irreconcilable propositions, each apparently self-evident, namely that South Africa is an independent state, and that the eradication of apartheid represents the culmination of the struggle to free Africa from colonial domination. Put in terms of South Africa's internal situation, a struggle essentially anti-colonial in origin and character is taking place in a country that has long ceased to be a colony.

In this context one may ask how to characterize this struggle. Is it an independence struggle, a national liberation struggle, or a struggle for democracy and civil rights? The question is a false one. It is necessary to grasp the concept of the *sovereignty of South Africa's people* before the themes of independence, national liberation, democracy, and civil rights fall into place.

South Africa has certain essential characteristics of independent statehood, but lacks the fundamental one: the internal co-existence of statehood and sovereignty. The mere existence of a territory, population, and a government exercising a degree of effective control is not enough.

The South African *state* is independent in the sense that it is not subject to the legal control of any other state, but the *people* are not independent inasmuch as they lack sovereignty. A state which reserves its sovereignty for a small racially-constituted minority, which negates the legal personality of the great majority on the grounds of indigenous origin, which deprives them constitutionally of citizenship rights, which leaves them without nationality, and subjects them to massive racial discrimination, cannot claim to be an 'independent state' in the full meaning of the term.

The exclusion of the majority from national sovereignty was structured through the bantustan policy, expressly designed to exclude people from the national polity under the guise of granting them independence in separate tribal states.

Self-determination, by its intrinsic nature, can never be endowed, even less imposed. It arises out of the determination of a nation or a people to attain independent statehood for themselves. The bantustans were designed in Pretoria to frustrate, not concede the national demands of the majority.

South Africa is not just one among many states in which people have no effective say in government; nor is it merely one of the many states where racism is practised. South Africa is an explicitly racist state, in which racist domination is built into the legal order as pervasively as colonial domination was built into the now dismantled Empires.

If, by certain criteria, South Africa is already described as an independent state, it is one in which the majority of the people have never enjoyed independence. Until the independence granted by Britain in 1910 to the white minority is extended to the whole population throughout the country, South Africa cannot be treated as an independent state in the full sense of the word. Its independence is inchoate, and will only be complete when sovereign power is exercised by the people as a whole.

The incomplete nature of South Africa's independence has a direct bearing on the next question, already described as a false one.

Should the anti-apartheid struggle be characterized as one for independence, national liberation, democracy, or civil rights?

International law, like nature, abhors a vacuum. The negation of the rights of the South African majority on the grounds of their national origin is the fundamental characteristic of apartheid from which all other features of the system flow. At the centre of apartheid lies the destruction of African independence and the usurpation of African land. In legislative terms, the foundations were the bloc of statutes restricting the land, rights, labour, movement, and residence of the African people. Coupled with the exclusion of the African people from the constitution, these laws constituted the foundation of apartheid, making the South African statute book the strongest proof that apartheid's central feature is the denial of the national rights of the African people.

Thus the struggle against apartheid presents itself as the culmination of the process of freeing Africa from foreign domination and liquidating the last relics of overt colonial conquest on the continent. It is true that the form of 'decolonization' differs in that the colonizers settled permanently in the colonized territory and established a state that was independent of the states from which they had come. An important legal consequence of this is that post-apartheid independence cannot mean secession and the creation of a separate state, but rather implies the elimination of the internal structures of domination which make the majority rightless in the land of their birth. It also implies that there is no other country to which members of the dominant minority owe allegiance or to which they could be expected to remove themselves once full local independence is achieved. They lose their 'right' to oppress the majority, but not their right to be full citizens in the land of their birth.

In South Africa, therefore, the form self-determination will take is the destruction of the barriers which exclude the majority of people from national sovereignty. Franchise rights on a basis of complete equality (one person, one vote on a common voters roll throughout the country) become the concrete political expression of the achievement of 'independence'. This coincides with the fundamental democratic notion that the people shall govern, that government is based not simply on the consent of the people, but on their will.

Thus national liberation — and through it, full independence — will be achieved in South Africa by the creation of a democratic state that is non-racial in its constitution and anti-racist in its activities. It follows that it would be wrong to attempt to define the struggle against apartheid as being either for national liberation or for democracy. It is for both. National liberation is the content, democracy the form.

A democratic state will replace a race-supremacist state not simply because democracy is good in itself, but because it is the only means of redressing in a fair and orderly way the great historic injustices consequent upon invasion, conquest, and domination, and subsequently institutionalized by the network of apartheid laws. Genuine popular sovereignty, which is at the heart of democracy, therefore presupposes far more than mere incorporation, step by step, into the existing political order. It presupposes restoration of usurped land and wealth, an end to national humiliation in all its forms, and an affirmation of the culture and personality of all the people.

The fundamental tasks of the democratic state will be to achieve those goals. Precisely how this should best be done is a matter which belongs to the new sovereignty. Thus, for example, the restoration of land could take many forms, reflecting many different philosophies. The land could be parcelled out to peasants; alternatively, the objective might be achieved by establishing agri-businesses with a non-racial shareholding, or co-operatives, or state farms, or joint ventures with state participation. There could be an infinite mixture of all these forms, even others conceived to meet the occasion. These are issues which the people of South Africa should be free to settle for themselves in a democratic and sensible way. That is what free debate, elections and Parliament are for.

It is in this context that the question of civil rights in South Africa must be viewed. If it is wrong to see the struggle for democracy as an alternative to the struggle for national liberation, it is even less correct to oppose the struggle for self-determination with the struggle for civil rights.

In South Africa, civil rights for all can only be a reality when the country is governed by the people as a whole and not by a racial minority. The fundamental question is not who can ride in railway carriages or sit on park benches or play in sports teams, important though these matters are, but who *decides* who can do these things. Who are empowered? To whom does the country belong?

To stress that apartheid is as deeply structured and as totally condemnable as slavery and colonialism is not to trivialize humanitarian criticism of its detailed aspects, nor to suggest that the struggle for civil rights is unimportant. The objective of a document such as the Freedom Charter was precisely to create the conditions in which the people of South Africa could enjoy full civil rights.

Looking to the future, our objective can never be to replace one form of dictatorial or authoritarian rule with another. Not even the most enlightened leadership can guarantee that human rights are respected at all times and in all cases. The work of the human rights activists, of community and other non-governmental organizations, never ends; progressive governments welcome them, even if they irritate at times.

The anti-apartheid cause is accordingly not well served by attempts to create either/or formulae for characterizing the nature of the struggle. What exists in reality is a single popular struggle taking a variety of forms, rooted in a wide range of democratic and patriotic forces with the goal of destroying the apartheid state and replacing it with a democratic state that is non-racial in character and anti-racist in programme. In the process of destroying apartheid and reconstructing South Africa, sovereignty is the essence, national liberation the substance, democracy the form, and human rights the goal.

Appendix 2
The ANC's Constitutional Guidelines for a democratic South Africa — proposed amendments after seminar on gender

The Freedom Charter, adopted in 1955 by the Congress of the People at Kliptown near Johannesburg, was the first systematic statement in the history of our country of the political and constitutional vision of a free, democratic, and non-racial South Africa.

The Freedom Charter today remains unique as the only South African document of its kind that adheres firmly to democratic principles as accepted throughout the world. Amongst South Africans it has become by far the most widely accepted programme for a post-apartheid country. We are now approaching the stage where the Freedom Charter must be converted from a vision for the future into a constitutional reality.

We in the African National Congress submit to the people of South Africa, and to all those throughout the world who wish to see an end to apartheid, our basic guidelines for the foundations of government in a post-apartheid South Africa. Extensive and democratic debate on these guidelines will mobilize the widest sections of our population to achieve agreement on how to put an end to the tyranny and oppression under which our people live, thus enabling them to lead normal and decent lives as free citizens in a free country.

The immediate aim is to create a just and democratic society that will sweep away the centuries-old legacy of colonial conquest and white domination, and abolish all laws imposing racial oppression and discrimination. The removal of discriminatory laws and eradication of all vestiges of the illegitimate regime are, however, not enough; the structures and the institutions of apartheid must be dismantled and be replaced by democratic ones. Steps must be taken to ensure that apartheid ideas and practices are not permitted to appear in old forms or new.

In addition, the effects of centuries of racial domination and inequality must be overcome by constitutional provisions for corrective action which guarantee a rapid and irreversible redistribution of wealth and opening of facilities to all. The constitution must also promote the habits of non-racial and non-sexist thinking, the practice of anti-racist behaviour, and the acquisition of genuinely shared patriotic consciousness.

The constitution must give firm protection to the fundamental human rights of all citizens. There shall be equal rights for all individuals, irrespective of race, colour, sex, or creed. In addition, it requires the entrenching of equal cultural, linguistic, and religious rights for all. *Special attention has to be paid to combating sexism, which is even more ancient and as pervasive as racism.*[*]

Under the conditions of contemporary South Africa eighty-seven per cent of the land and ninety-five per cent of the instruments of production are in the hands of the ruling class, which is solely drawn from the white community. It follows, therefore, that constitutional protection for group rights would perpetuate the status quo and would mean that the mass of the people would continue to be constitutionally trapped in poverty and remain outsiders in the land of their birth.

Finally, success of the constitution will be, to a large extent, determined by the degree to which it promotes conditions for the active involvement of all sectors of the population at all levels in government and in economic and cultural life. Bearing these fundamental objectives in mind, we declare that the elimination of apartheid and the creation of a truly just and democratic South Africa requires a constitution based on the following principles:

The state:

a) South Africa shall be an independent, unitary, democratic, non-racial and non-sexist state, *based on the principle of equal rights for all.*

---

* The italics represent amendments proposed at a seminar jointly organized by the Women's Section and the Constitutional Committee of the ANC. The changes were adopted after four days of discussion informed by comments and papers received from inside and outside South Africa. About seventy persons, roughly two-thirds women, one-third men, attended. The italicized amendments correspond to the author's record of the proceedings and should not be regarded as the official text.

b) i. Sovereignty shall belong to the people as a whole and shall be exercised through one central legislature, executive, and administration.
    ii. Provision shall be made for the delegation of the powers of the central authority to subordinate administrative units for purposes of more efficient administration and democratic participation.
c) The institution of hereditary rulers, chiefs, *and chieftainesses*, shall be transformed to serve the interests of the people as a whole in conformity with the democratic principles embodied in the constitution.
d) All organs of government including justice, security, and armed forces shall be representative of the people as a whole, *men and women*, democratic in their structure and functioning, and dedicated to defending the principles of the constitution.

Franchise

e) In the exercise of their sovereignty, *all men and women* shall have the right to vote under a system of universal suffrage based on the principle of one person, one vote.
f) Every voter shall have the right to stand for election and be elected to all legislative bodies.

National identity

g) It shall be state policy to promote the growth of a single national identity and loyalty binding on all South Africans. At the same time, the state shall recognize the linguistic and cultural diversity of the people and provide facilities for free linguistic and cultural development. *Such cultural diversity shall not be the basis for discrimination.*

A Bill of Rights and affirmative action

h) The constitution shall include a Bill of Rights based on the Freedom Charter. Such a Bill of Rights shall guarantee the fundamental human rights of all citizens irrespective of race, colour, sex, or creed, and shall provide appropriate mechanisms for their enforcement.
i) The state and all social institutions shall be under a constitutional duty to eradicate race discrimination in all its forms.
i) **bis** *The state and all social institutions shall be under a constitutional duty to work towards the rapid elimination of inequality based on gender and to combat sexism in all its forms.*

j) The state and all social institutions shall be under a constitutional duty to take active steps to eradicate, speedily, the economic and social inequalities produced by racial discrimination.
k) The advocacy or practice of racism, fascism, nazism, or the incitement of ethnic or regional exclusiveness or hatred shall be outlawed.
l) Subject to clauses (i) and (k) above, the democratic state shall guarantee the basic rights and freedoms, such as freedom of association, expression, thought, worship, and the press. Furthermore, the state shall have the duty to protect the right to work, and guarantee education and social security.
m) All parties which conform to the provisions of paragraphs (i) to (k) shall have the legal right to exist and to take part in the political life of the country.
m) **bis** *The basic rights and freedoms set out above shall be enforceable through the courts and the principle of ensuring equal access to the legal system shall be followed.*

Economy

n) The state shall ensure that the entire economy serves the interests and well-being of all sections of the population.
o) The state shall have the right to determine the general context in which economic life takes place and define and limit the rights and obligations attaching to the ownership and use of productive capacity.
p) The private sector of the economy shall be obliged to co-operate with the state in realizing the objectives of the Freedom Charter in promoting social well-being.
q) The economy shall be a mixed one, with a public sector, a private sector, a co-operative sector, and a small-scale family sector.
r) Co-operative forms of economic enterprise, village industries, and small-scale family activities shall be supported by the state.
s) The state shall promote the acquisition of managerial, technical, and scientific skills among all sections of the population, especially the blacks, *and shall take special steps to remove the barriers to women participating fully in economic life.*
t) Property for personal use and consumption shall be constitutionally protected.

Land

u) The state shall devise and implement a Land Reform Programme that will include and address the following issues:
    i. Abolition of all racial and gender-based restrictions on ownership and use of land.

Constitutional Guidelines for a democratic South Africa

    ii. Implementation of land reforms in conformity with the principle of affirmative action, taking into account the status of victims of forced removals.

Workers

v) A charter protecting workers' trade union rights, especially the right to strike and collective bargaining, shall be incorporated into the constitution.

Women and men

w) *A charter of gender rights shall be incorporated into the constitution guaranteeing* equal rights *between men and women* in all spheres of public and private life and requiring the state *and social institutions* to take affirmative action to eliminate inequalities, discrimination, and *abusive behaviour based on gender.*

The family

x) The family, parenthood, and *equal rights within the family* shall be protected.

Children's rights

y) *The principles of the International Convention on the* Rights of the Child *shall receive constitutional respect.*

International

z) South Africa shall be a non-aligned state committed to the principles of the Charter of the Organization of African Unity and the Charter of the United Nations and to the achievements of national liberation, world peace, and disarmament.

Index

abortion 56, 59, 61, 68–9
accommodation rights 126
affirmative action 12–13, 19–21, 23, 28, 29, 37, 169–73, 183, 191, 199–200
African culture 180–1
    and the environment 142
African peasant and new constitution 28–9
African women 54–5; *see also* gender rights
Africans
    destruction of independence of 194
    legal profession and 91, 98
Afrikaans language 27–8, 162, 180, 181
Afrikaans literature 181
Afrikaner businessman: constitutional relationship with African peasant 26–9
Afrikaners
    and environment 142–3
    and residential exclusivity 28
    freedom struggle 187
    group rights 155
    see also group rights
Afrikaner culture 142, 181
aged, rights of the 26
agri-businesses, non-racial 196
agricultural debt 133
agricultural land *see* land
agricultural workers *see* farmworkers
AIDS 69
ANC 177–8
    and Bill of Rights 37
    founding of 152
    non-African members 178
ANC Constitutional Guidelines 37, 138, 197–201
    culture and 175–83
    proposed amendments after seminar on gender 197–201
anthem 43, 45, 178
anti-apartheid struggle 188, 193–6
    effect on children 86
anti-trust legislation 167
apartheid
    and African women 55, 65
    and children's rights 79–80
    and conservation 140–1
    and family life 64–70 *passim*
    and majority rights 194
    and the legal system 87, 90, 94, 96–7
    democratic 4
    dismantling 197–8
    freedom as reverse of 184
    hidden 4
    multiracial 3–4
    open 2
    privatized 157–8
    reformed 2–3
    residential 28, 158, 200
army 10, 48, 88
Army and Security Commission 21
art and politics 175–7, 183
Asmal, Kader 9
assessor system 95
atheism 44, 49
authoritarianism vs. constitutionalism 32

Bantu Education law 88
Bantu law 71
Bantustans 2, 3–4, 36, 136, 152, 194
Bill of Rights 5–31, 159, 172, 173, 190, 191
    affirmative action 12–13, 19–21, 191, 199–200
    and land rights 108, 109, 137
    and majority rule 33–4
    constitutional guidelines 199–200
    constructing 16–17
    federalism and 151
    functions of 33–4, 37, 42, 189
    fundamental principles 41
    implications for culture 182
Bill of Rights culture 32–42
Birley, Sir Robert 81
birth control 67–8
'blue rights' 144
bohadi 54, 70
businessman, Afrikaner: implications of new constitution for 26–8
Buthelezi, Chief Gatsha 3

Cape Law Society 97
capital punishment 94, 99
caretaker administration 156
Ceteswayo 180
child abuse 79, 88
child care 29, 62, 69, 83, 87
child welfare 80
childhood, the right to 85
children
    apartheid and 79–80
    effect of conflict on 86

social programmes 87–9
children's ombudsman 88–9
children's organizations 84, 88
children's rights 26, 79–89
　apartheid and 79–80
　charter of 79, 87, 88–9
　constitutional guidelines 201
　corrective action 88
　strategy of guaranteeing 87–9
Christianity *see* religion
church and state, relationship between 45–6
church groups 103
churches 10, 17
　African 47, 49, 72
citizenship 191
civil law 93
civil procedure 93
civil rights 196; *see also* Bill of Rights; human rights
colonialism and independence 193, 194, 195
commercial law 92, 93
common law marriage 72
communal farming 115, 135
community and children's rights 88
community courts 78, 102–3
community involvement 153
community legal services 97
community organizations 103, 188
compensation for property 11, 105, 128 130–1, 134, 138, 157, 166
compound system and family 66, 67
concurrent interest in farmland 94, 105
confession, forcing of 10
confiscation of land and goods 28
Congress of the People 16
conscience, the right to a 43
conscientious objection 47–8
consensus 38
conservation rights 139–48, 146, 190
conceptional representation 156
constitution
　and Bill of Rights 32–7
　and entrenched guarantees 157–8, 190–1
　anti-apartheid foundation of 184–5
　basic principles of 159
　children's rights 86
　democratic character of 186–7
　functions 33
　making of 30–31

women's role 57, 63
Constitution of the Union of SA 34
Constitutional Guidelines of the ANC 37–8, 138, 175–83, 197–201
constitutional position of whites 26–8, 149–74
constitutional schemes 1–5
constitutional vs. electoral issues 33
constitutionalism vs. authoritarianism 32
consumer rights 94, 142
contraception 68–9
contract, freedom to 114–6
contract, inviolability of 158
contract law 92, 94
contractual rights and land rights 121
co-operative farming 29, 105, 115, 135, 196
corporal punishment 107, 178
courts 21, 35, 77, 87, 200
　access to 190, 191
　affirmative action 21
　Africanization 95, 100–3
　and the gender question 55
　informal 102
　plural or unitary systems 100–101
creed (word) 44
criminal law 93, 94
criminal procedure 93
cultural rights 28, 82, 121–2, 154, 156, 163, 179, 191, 198, 199
　and political rights 160–3, 179
　and the environment 142
culture
　African 142, 162, 180–1
　Afrikaner 142–3, 162, 181
　ANC Constitutional Guidelines 175–83
　as an instrument of struggle 175–8
customary law 70–71

Daily Mail 52
Defiance of Unjust Laws Campaign 98
delict 94
democracy
　Bill of Rights and 17–19
　goals 195–6
　inside the family 65
　see also non-racial democracy
democratic constitution 186–7
detention without trial 98
discrimination 163–4
　eradication of unequalities of 200
　guarantee against 24, 154, 160–4, 191

divorce 75, 80, 90, 101, 102
domestic servants 55
due process 7, 34, 98, 138
Dugard, John 95
Dumile 177
Dutch Reformed Church 27
ecological duties 190
economic control 167
economic dimension of land ownership 131–6
economic enterprise 200
economic policy 165–8
economic rights 121–2, 145, 200
economy
    affirmative action 200
    constitutional guidelines 200
education 169
    and liberation 86
    unequal access to 69
education charter 17
education rights 27–8, 81–2, 122, 200
    affirmative action 81, 122
elections 191, 199
electoral vs. constitutional issues 33
employment rights 123, 168, 200
    see also farmworkers; work; workers
employment conditions 123, 132, 145
employment practices and family life 67
enforcement of rights and freedoms 200
    see also affirmative action
English language 91, 156
entrenched guarantees 191
entrenchment of privilege 10, 15, 157–8
entrepreneurial activity 167
environmental code 147
environmental law 143
environmental ombudsman 147
environmental rights 139–48, 190
equal opportunities 191
equal opportunities commission 62
equal rights 160–4
    guarantees of 154, 191
equal rights clause 29
equality and identity 161–2
ethnic exclusiveness 200
European Convention of Human Rights 166
eviction 122, 123, 125, 130
exclusiveness 28, 158 , 200, 201
expression, freedom of 50–2
expropriated land, return of 129
expropriation 128, 168

family 56, 64–78, 88, 102, 103
family courts 55, 77–8
family law 54, 65, 66–76, 69, 92, 93
    African 54, 100–3
    mechanisms for implementing 77–8
    pluralist 75, 100–2
family life, apartheid and 55, 64–70
family planning 67–8
family rights 55, 56, 123, 201
farmers
    parliamentary representation 107
    rights 28 (*see also* land rights)
    subsidization 107, 133
farmers, black 115
farmers, white 107
farmworkers' rights 122–3, 132
    women 55, 123, 130
federalism 3–4, 151–3
fellowship: right to 84
feminism 35, 61
fertility control 67–8
fiduciary rights 144
first generation of rights 7, 21, 144, 190
Fischer, Bram 98
forced removals: return 10, 129, 131, 201
franchise 106, 118, 151, 155, 199
free enterprise 114, 165, 166
freedom 184–91
Freedom Charter 4–5, 13, 16, 17, 33, 111, 159, 176, 185, 188, 196, 199
    and constitutional guidelines 197
    and cultural rights 156
    and rural rights 138
freedom of assembly 9, 159, 191
freedom of association 25, 158, 200
freedom of organization 130
freedom of speech 9, 50–2, 130, 159, 190, 191
freedoms, basic 200

Gandhi, M. K. 98
gender rights 26, 29, 53–63, 123, 130, 62, 199–201
    charter 61–3 *passim*
    constitutional guidelines 198–201
    right of preference 164–5
gender rights council 62
gender stereotyping 56, 60, 61
government, system of 159, 191, 199
'green rights' 145
group areas 10, 11, 28, 108, 139, 173
group libel 50

group representation 156
group rights 149, 150, 152, 154, 161, 198
   and the Bill of Rights 37, 39–40
   guarantee of 24–5
   see also minority rights
guardianship rights 144
Gwanga, Jonas 177

handicapped, rights of the 26
Head of State 4, 5
health rights 60, 62, 69, 80, 87–8
health services 169
hereditary rulers 199
history, respect for 28
holidays, national 28
homelands 2, 3–4, 36, 136, 152, 194
homosexuality 165
housing 67, 83
human rights 7–9, 18, 21, 35, 138–48,
   190, 198, 200
     and entrenchment of privileges 157
     and property rights 116–18, 119–27,
     131–2, 138
     re-allocation of 7
     United Nations charter of 7
     Universal Declaration of 6, 7
     universalization of 92
     see also Bill of Rights; civil rights;
     gender rights *etc*

Ibrahim, Abdullah 177
identity 160–3
illegitimacy 88
imagination, children's right to 82
independent statehood 193–6
individual rights, guarantee of 24
infertility 69
information, right to 52, 88, 145
inheritance, right to 88, 123
Inkatha 176
International Convention on the Rights
   of the Child 87, 201
International Court of Justice 92

Jehovah's Witnesses 47
Jews 44, 101
job security 157
Johannesburg Stock Exchange 167
judicial review 35, 191
judiciary 35, 36, 39, 77–8, 88, 91, 94–5,
   100, 159, 191
jury system 95

Kotze, Judge 35

Krause, Judge 99
Kruger, Paul 35
KwaZulu-Natal Indaba 3

labour law 169
labour tenancy 101, 114, 116, 125
land, access to 67, 118, 134
land, dispossession of 104–6
land, redistribution of 10–12, 29–30, 105,
   108–38 *passim*
land, usurpation of 194
land abuse 94
Land Act 108, 114, 137, 172
Land Bank 107, 133, 169
land claims 108, 118, 128–30, 138
Land Commission 21
land court 130
land law 94, 106–27 *passim*, 134
   see also property law
land ownership 107, 123–7, 128–31,
   132–3
     constitutional guidelines 200–1
     criteria 128–9, 130
     shared 128, 133, 135
Land Reform Programme 200–1
land rights 10–11, 104–38, 200
     and constitutional rights 119–27
     de-racializing 108–11
     different rights 108, 118
     integrity 147
     shared 116, 124–7, 128
     women's rights 123
     see also land, redistribution of
land tax 134
language 27–8, 77, 156, 162, 180, 181
   gender-biased use of 56
language policy 156, 162
language rights 25, 27, 28, 156, 162–3,
   179, 191, 198, 199
law 90–103
   African 100–3
   apartheid concept of 87
   codification 90, 94
   English 90
   Mozambican 90, 93
   political functions 19
   private 90
   public 90, 93
   role in new nation 87
   Roman Dutch 90–103 *passim*
   South African 90–103
   Zimbabwean 93
   see also civil law; contract law *etc*

law societies 91, 97–8
lawyers, role of 103
leadership 182–3
lease 106, 125
legal culture, universal 92
legal profession 94–9, 103
  racism in 97–8
legal services 97 *bis*
legal system 90, 92, 93
  access to 96–7, 190, 191
  African component 100–3
  and human rights 21, 200
  apartheid and 90, 94, 96–7
  plural or unitary 100–101
legal training 96
linguistic rights *see* language rights
lobola 66, 70, 71, 73, 75, 101–2
local government 153, 191
location system and family life 66
lotteries 45
love and the freedom struggle 176
Luthuli, Chief Albert 107

Magna Carta 9
majorities and minorities 21–6
majority rule 6–7, 17–18, 33–4, 159
Makeba, Miriam 177
makgotla 102
Mandela, Nelson 16, 92, 98
Mangena, Alfred 91–2
marriage 45, 54–5,
  matrimonial law 72–6 *passim*, 94, 95–6, 101–2, 103
  traditional 70, 101–2
Masekela, Hugh 177
Mass Democratic Movement 38
master and servants law 94
maternity benefits 29
Mbeki, Thabo 52
media 51–2, 56
  see also press
migrant labour 55, 66, 67, 83, 152
military service 47
mineral rights and compensation 134–5
minorities and majorities 21–6
minority fears 159–60
minority rights 6–7, 15, 145, 153–8
  protection of 15, 153–8
monopoly control 167
monuments 28
mother care 69, 87
Mozambique 90, 93, 103
municipal services 169

music 177
Muslims 44, 71–2, 74, 75
Muzorewa, Bishop 31

name, right to a 66, 79, 88
Natal 3
National Education Crisis Campaign 17
national identity 199
national liberation 194–6
national unity and cultural diversity 179
nationalization of land law 108, 134
Native Labour Regulation Act 66
Native Lands Act 11
Native Urban Areas Act 66
negotiations 15–16, 38
neighbourhood organizations 67
New Nation 52
Nkosi Sikelel' i Afrika 43, 45, 178
non-racial democracy 4–5, 186
  position of whites in 159–74
official languages 34
one person, one vote 155, 190, 195, 199
Organization of African Unity 201
Own Affairs system 3
ownership 107, 123–7, 166, 200
  shared 124–7, 166
  see also property

pacifism 47–8
parenthood 201
parents
  children's rights 79, 88
  single 55, 56, 70
Parliament 3, 36, 191
party system 159, 191, 200
pass laws 55, 66, 112
patriarchy 53, 54, 55, 56, 64
peace, the right to 145
peasants, granting of land to 196
pensions 67, 157
people's courts 102
people's rights 145
place names 28
pluralism 156, 159, 163
police 10, 88
political parties 159, 191, 200
political pluralism 159, 163
political rights 200
  and cultural rights 160–3
politics
  and art 175–7, 183
  and law 19
poverty and environment 141

Index

pre-school feeding 87
press 51–2, 78, 200
Pringle, Thomas 143
private sector of the economy 200
privatization 165
privatized apartheid 157
privilege, entrenchment of 10, 15, 157–8
productivity and land claims 133–4
property 165–8, 200
    redistribution *see* redistribution of property
    see also land
property law 106–11
property rights 18, 27, 28, 104–38, 165–8, 200
public domain and private rights 163–4
public law 90, 93
Public Service Commission 21
punishment 10

race classification, voluntary 154–5
racial discrimination *see* discrimination
racial exclusiveness 28, 158, 200, 201
racism 194, 200
    and free speech 50–2
    in the legal profession 91, 97–8
Rand Daily Mail 51
reactionaries 68
'red rights' 144–5
redistribution of property 10–12, 29–30, 105–38 *passim*, 167–8, 191, 195–6, 198
regional government 191
religion 43–9
religious fundamentalism 49
religious leader activists 17
religious rights 25, 27, 43–9, 154, 191
    charter 17, 46–7
religious-cultural association 25
remedies and rights 146
rent control 126
residential exclusiveness 28, 158, 200
residential rights 28, 122
resources 8, 10, 145, 157, 166
restoration of land *see* redistribution of property
revolutionary seizure of power 30
revolutions 1
rights, basic 200
Roman Dutch law 90–103 *passim*, 108, 123, 125
Roman Law 92
Rose-Innes (Judge) 94, 99
Rule of Law 8, 91, 159, 173, 190

and land rights 109–10, 119, 120–1
rural areas, rehabilitation of 67
rural rights 118, 119–27, 131–2, 138
    charter 138

school-feeding 69, 87
schooling 69; *see also* education
Schreiner, Olive 143, 180
second generation rights 7, 18, 21, 144–5, 190
sectional title 126
Security and Army Commission 21
self-determination 145, 195
separation of powers 159, 191
Serote, Wally 180
sex education 69
sexism 57, 60, 198, 199; *see also* gender rights
sexual abuse and harassment 55, 56, 62, 123
sexual preference, right of 165
Shaka 180
share-cropping 106–7, 114, 116, 125
single parents 55, 56, 70
slavery and family life 65–6
Social and Economic Rights Commission 21
social programmes 87–8
social security 67, 200
socialism 169
socio-cultural rights 25, 27
socio-economic rights 18, 121–2, 145; *see also* second generation rights
solidarity, rights of 145
Solomon, 43, 94
SOMAFCO 178
South African Law Commission 37
sovereignty 106–10, 193–6, 199
state, the and religion 45–6
    children's claims against 79–80
    constitutional guidelines 198–9
state farms 105, 196
state of emergency 84, 98
state purchase of land 29
statehood 193
status quo, maintenance of 157–8, 198
statutory presumptions 36
strike, right to 122, 168, 201
succession, laws of 107
suffrage, universal 159, 199
    see also franchise
tax system and family life 66, 67
taxation, native 112

tenancy, protected 126
tenant farmers' rights 29–30
theocracy 44, 46
third generation rights 7, 18–19, 21, 139–48, 190
time-sharing ownership 126
Torch Commando 187
trade unions 17, 48, 67, 103, 168, 187, 201
    agricultural 122, 130
transitional arrangements 12, 30–1, 38, 156, 157, 166
Transvaal Law Society 91, 98
tribalism 51
Tutu, *Archbishop* 43
tyranny and majority rule 34

Union of South Africa Act 34
United Nations 92, 201
    convention on children's rights 87
United States Bill of Rights 9–10
Universal Declaration of Human Rights 6, 7, 9, 33
'Urban Blacks' and reformed apartheid 3
urban land, rights to 128
usufruct 124, 125, 129
utilities, the right to 141

vested interests 157, 172
violence against blacks 10, 121
violence against children 81
violence against women 56, 59, 62, 123
vote, right to *see* franchise
voters' role 34, 155, 159, 195

wage system and family life 67
Weber, Max 109
Weekly Mail 52
welfare support 56
white farming 133–5
whites
    constitutional position 149–74
    privilege 10, 15, 157–8
    in the anti-apartheid struggle 181–2
widows 54, 56
women
    African 54–5
    employment 59, 62
    legal profession 77, 78, 97–8, 103
    role in formulating constitution 57
    unmarried 55
    violence against 56, 59, 62
women's organizations 17, 57, 78
women's rights *see* gender rights

work, right to 200
work conditions 123, 132, 145
work environment 147–8, 168
workers
    and the gender question 55, 123, 130
    charter 17, 168
workers' rights 26, 145, 168–9, 201
    see also farmworkers' rights
worth, the right to 83–4

Zimbabwean law 93, 136
Zulu culture 180